12-19-95

D1622248

GENETICS AND MEDICINE
in the
UNITED STATES
1800 to 1922

GENETICS AND
MEDICINE
in the
UNITED STATES
1800 to 1922

ALAN R. RUSHTON, M.D., Ph.D.
Associate Clinical Professor, Department of Pediatrics
Robert Wood Johnson Medical School, Rutgers University
Piscataway, New Jersey

THE JOHNS HOPKINS UNIVERSITY PRESS

Baltimore and London

© 1994 The Johns Hopkins University Press
All rights reserved
Printed in the United States of America on acid-free paper

The Johns Hopkins University Press
2715 North Charles Street, Baltimore, Maryland 21218-4319
The Johns Hopkins Press Ltd., London

ISBN 0-8018-4781-8

Library of Congress Cataloging-in-Publication Data
will be found at the end of this book.
A catalog record for this book is available from the British Library.

To Nancy

CONTENTS

PREFACE

Genetics is a young science. The rediscovery of Gregor Mendel's work in 1900 launched the intensive investigation of heredity in plants, animals, and eventually humans which has resulted in the detailed understanding of the genome and its actions available today. General histories of this intellectual process have begun to appear over the past two decades. *The Gene: A Critical History* by E. Carlson (1966) was a pioneering attempt to explain how this fundamental concept in heredity came about. *A History of Genetics* by A. H. Sturtevant (1965) and *History of Genetics* by H. Stubbe (1972) cover the major discoveries that led to the modern science of genetics. Specific aspects of Mendelian genetics have fostered the leading paradigms for research work in this entire field of endeavor. Its evolution has been considered in several recent works. *Origins of Mendelism* by R. Olby (1985), *The Growth of Biologic Thought* by E. Mayr (1982), *The Mendelian Revolution* by P. J. Bowler (1989), and *Theory Change in Science: Strategies from Mendelian Genetics* by L. Darden (1991) examine the biologic and sociologic factors that resulted in the success of this theoretic model for heredity.

The history of human genetics has been strangely neglected by historians of science. Two brief articles from the 1970s touched upon this matter (McKusick, 1975; Reed, 1979), but no comprehensive work has yet examined the development of human heredity in the United States or Great Britain. *The Foundations of Human Genetics* by K. Dronamraju (1989) outlines the most recent work in this area from 1930 to 1989, but no investigation has studied the role of human heredity during the formative years of Mendelian genetics, from 1900 to 1920.

The political offshoot of human genetics—eugenics, or social betterment through selective breeding—has provided a fertile area for detailed historical investigation. *Genetics and American Society* by K. Ludmerer (1972), *Eugenics: Hereditarian Attitudes in American Thought* by M. Haller (1984), and *In the Name of Eugenics* by D. Kevles (1985) examine the application of human genetics to U.S. social and political life in the nineteenth and twentieth-centuries.

A reviewer of an early draft of this book commented that it sounded like history written by a scientist. Indeed, that is the case. As I reviewed the historical sources on genetics in U.S. medicine, I attempted to think about history as a medical scientist. This work has focused on medical genetics in an historical context. Although I have tried to place my medical sources in the social and intellectual context of the day, this is not primarily a social or political history. And although of necessity I have considered many aspects of human heredity, the primary focus of this work is that portion of human heredity called medical genetics. This has to do with understanding how hereditary factors produce disease symptoms in human beings.

The primary sources for this historical study were identified from the major indices of the late nineteenth and early twentieth centuries. *The Index Catalogue of the Library of the Surgeon General's Office* lists five pages of references on heredity and disease published before 1885 (Billings, 1885). The most important single source was the *Index Medicus,* which began publication in Washington, D.C., in 1879, ceased in 1899, and resumed in 1903. The interlude is covered by *Bibliographia Medica,* published in Paris from 1900 to 1902. These two indices do not have annual compilations. The availability of *Index Medicus* on microfilm (Princeton Microfilm Corporation, Princeton, N.J.) enabled a "brute strength" approach, which involved reading the entire contents of each annual volume to locate articles from the literature whose titles appeared to be relevant to this investigation.

My interest in the history of genetics developed as the result of undergraduate courses at Earlham College. Professors Jerry Woolpy and Jerry Bakker taught science courses from an historical perspective and encouraged analysis of competing ideas within professional disciplines.

I evaluated most of the scientific articles while I was a faculty member at Princeton University. The Firestone Library contains a complete collection of the U.S. biological publications from this era. The library at the New York Academy of Medicine provided most of the medical sources in this work, a magnificent historical collection that included almost all the references that I requested. I wish to thank the library staff at the academy for their efforts in locating the thousands of references for my study. More recent articles were obtained through the kindness of medical librarians Joyce White and Jeanne Dutka at Hunterdon Medical Center. I appreciate their work on interlibrary loan to obtain these needed materials.

My medical colleagues at the Department of Pediatrics of Hunterdon Medical Center also fostered this project by permitting me to use educational leave for the library research in this work. My family has been an important part of this project as well. My sons, Andrew and

Daniel, have learned about computers and printers in their dad's study. And my wife, Nancy Spencer Rushton, suffered piles of index cards and a preoccupied husband for several years. She carefully read the entire manuscript and made helpful editorial comments to improve the quality of my written work. I deeply appreciate her attention and support.

The editorial staff at the Johns Hopkins University Press has labored hard over this manuscript as well. Wendy Harris took the time to read successive drafts and then encouraged me to make the necessary revisions. Roberta Hughey has transformed the final draft into a more unified and readable entity.

This book is heavily referenced to encourage other scholars to investigate specific aspects of this topic in the future. I hope that my work will act as a useful index to the literature on heredity and disease for this period.

The research in the book shows that few U.S. physicians were involved in investigations in human genetics between 1900 and 1920. The situation in Great Britain appears to have been different. English physicians were better educated in the natural sciences (Peterson, 1978). The eminent physician Archibald Garrod collaborated with William Bateson on genetic studies of rare "inborn errors of metabolism" that resulted in human disease (Garrod, 1902). Karl Pearson, the director of the leading institution for human genetic research, the Galton Eugenic Laboratory in London, was interested in the medical aspects of human heredity and employed a physician investigator as early as 1907 (Farrall, 1985). A brief comparative study of genetics in the United States and Great Britain has already been published (Kevles, 1980). I plan a detailed analysis of the role of genetics in British medicine to 1920 as a sequel to the current study, which focuses on the scene in the United States.

GENETICS AND MEDICINE
in the
UNITED STATES
1800 to 1922

ONE

THE EARLY YEARS
1800 to 1860

At the beginning of the nineteenth century, there was no science of heredity. This is not to imply that questions of the hereditary transmission of particular traits from parent to child were of no concern. On the contrary, there was a lively interest in heredity in the day-to-day work of practitioners of animal husbandry as well as medicine. Breeders of domestic animals had long recognized that careful choice of mating pairs could modify many different physical traits. There was no agreement, however, on the relative roles played by heredity and local environment in the production of these variations in animal offspring (Russell, 1986). Physicians agreed that the same general rule held for human heredity: parents and their children were usually alike in most respects. Variations certainly did occur as external factors altered the correct functioning of this copying mechanism, which operated across the generations (Bowler, 1989). Attempts to understand the function of heredity in animals and humans at this time implied an interest in the causes of hereditary transmission of both normal and abnormal characters. But U.S. medical practitioners of the day had little time to consider such theoretical questions. Their primary function was to deal with the symptoms of their patients and to relieve suffering whenever possible.

VIGOROUS TREATMENT: U.S. MEDICAL PRACTICE, 1800 TO 1840

The major demand from the general public throughout the period from 1800 to 1840 was that a physician do something to cure a patient's symptoms. Any interest in causes of disease or in new and safer methods of treatment was secondary to relieving the symptoms of the suffering populace (Haller, 1981). The public expected vigorous therapy from the physician. And some specific treatments did appear to relieve the symptoms in certain cases: for fullness of the stomach, emetics; for diseases of the bowels, purgatives; for inflammatory diseases, bleeding; for intermittent fevers, Peruvian bark; for syphilis, mercury (Rothstein, 1972 who quotes Thomas Jefferson). But physicians rarely used only one treatment and then let nature attempt to heal the patient. They were, after all, paid to do something for the patient. They had to demonstrate their concern for patients' welfare by actively attacking symptoms (Rothstein, 1972). It was assumed that disorders with a similar symptom, such as fever, always had similar underlying causes; therefore, they should respond to similar therapies. Useful therapy was deemed to be anything that altered the pathological symptoms.

Bleeding was generally the initial therapy for a variety of conditions ranging from convulsions, concussion, and croup to smallpox, hernia, and pneumonia. It appeared to be somewhat less detrimental to the patient's general health than the prolonged use of harsh medications. The general aim of the physician was to remove as much blood as possible from the patient, and bleeding was often dramatically effective in relieving some of the symptoms. Within twenty minutes, copious perspiration would appear on the patient's face, and the fever and delirium would disappear. Then sedatives, nourishment, and time were prescribed to ensure complete recovery (Rothstein, 1972; Haller, 1981).

A second form of therapy was popularized by Benjamin Rush, the most prominent physician of this era. After an extensive medical education in Philadelphia, Edinburgh, and London, he was appointed professor of the theory and practice of physick at the University of Pennsylvania Medical School. His lectures and writings greatly influenced an entire generation of medical practitioners in the United States (L. King, 1991). He used cathartics such as calomel (chloride of mercury) with good results during the 1793 Philadelphia yellow fever epidemic, a treatment that appeared to be useful for all kinds of fever and inflammation and that was indicated for the removal of toxins from the teeth, feet, pancreas, kidneys, and skin (Haller, 1981).

Blistering was another popular method used to draw toxins from the body. Plaster was applied to the affected part to raise a blister, which was then broken to allow harmful material to drain from the body. The physician often intentionally irritated the remaining ulcer to increase the local production of pus, which drained the body of disease (Rothstein, 1972).

The public expected such vigorous therapy from physicians, for the United States was a young, vigorous country. One contemporary patient was reported to have remarked: "Now doctor, if you give me a dose, give me a big one" (Rothstein, 1972).

The demand for medical services in the United States at this time was increasing. Immigration and westward movement resulted in the expansion of existing urban centers and the establishment of new cities and towns on the frontier. The term *physician* applied to anyone who had an interest in this vocation—a farmer who could set a broken bone or a graduate of the most sophisticated European medical school of the day. Most U.S. physicians were the product of apprenticeships. Young men joined an established practitioner for four to seven years and paid an annual fee for instruction. The apprentice helped mix drugs and traveled with the physician as he visited patients, to learn the techniques for bleeding, blistering, pulling teeth, setting fractures, and draining abscesses. The best preceptors required that a student obtain a working knowledge of Latin, mathematics, and natural history as integral parts of his medical education. Textbooks from the doctor's personal library constituted reference material for the student to consult.

Those who could afford the expense would then study in Europe to supplement the practical training. Travel to London, Edinburgh, Paris, or Vienna allowed the young U.S. physicians to observe clinical work in large hospitals and eventually to obtain advanced degrees from local universities. Upon returning home, some of these young graduates attempted to organize medical schools to emulate the European model for clinical teaching. But this was a very slow process. By the time of the American Revolution, approximately 3,500 "physicians" engaged in practice in the colonies. Of these, 350 were graduates of European schools, while only 51 were the product of U.S. medical colleges (Haller, 1981).

By 1800 ten medical schools had been established in the United States. Candidates for admittance were supposed to have completed an apprenticeship and demonstrated competence in Latin and the natural sciences. A series of didactic lectures in anatomy and physiology, surgery, obstetrics, therapeutics, and chemistry was offered each year. Students relied on lecture notes for their basic study material because few medical textbooks were available in the United States. Attendance at two lecture sessions was required for graduation. The same lectures were offered each year so that the entire class would have adequate opportunity to grasp the material. Oral examinations then assessed the competence of each student.

The quality of the medical school graduates was low, for there were no recognized standards for the required apprenticeship program and the oral examinations in medical school were not difficult. Almost anyone who paid the lecture fees could obtain a medical degree. Rigorous study to obtain a detailed knowledge of medicine was unnecessary. The medical degree was not necessary either, because the practice of medicine was open to anyone who called himself "Doctor" (Haller, 1981).

One mechanism to improve medical education involved the establishment of professional journals to communicate new developments in practice. Shortly after the turn of the nineteenth century, several journals began publication in the eastern cities to disseminate information useful for the practitioner. *The Medical Repository*, the *Medical and Physical Journal*, and the *Medical Museum* were the earliest Philadelphia publications. *The Medical and Physical Register* was established in New York. *The Philadelphia Journal of Medical and Physical Sciences* became the *American Journal of Medical Sciences*, and the *New England Journal of Medicine and Surgery* eventually evolved into the *Boston Medical and Surgical Journal* (Packard, 1931). These early professional journals provided a forum for practitioners to share ideas on the diagnosis and care of the myriad of diseases they encountered throughout the young nation.

Out of this rough-and-ready approach to medical practice appeared in 1803 a publication by John Otto on heredity and disease that has been called "one of the earliest significant contributions to medical science by an American" (Kaufman, Galishoff, and Savitt, 1984).

HEREDITARY HEMOPHILIA AND BLINDNESS

Born in New Jersey in 1777, John C. Otto became a prominent Philadelphia physician in the first half of the nineteenth century. His father, grandfather, and great-grandfather had all been physicians. He graduated from the College of New Jersey (now Princeton University) in 1792, became an apprentice with Benjamin Rush in Philadelphia, and received the M.D. degree from the University of Pennsylvania in 1796. Otto developed an extensive medical practice in Philadelphia and followed Rush as physician to the Philadelphia Hospital, continuing in active practice until his death in 1844 (Krumbhaar, 1930; Packard, 1931).

Through his association with Rush, Otto became aware of several families of "bleeders" who had sought treatment from the prominent physician. But Rush declared that he was "not inquisitive enough for particulars" and recommended that Otto investigate this unusual pattern of illness.

Otto described in graphic detail the essential features of this condition:

> About seventy or eighty years ago, a woman by the name of Smith, settled in the vicinity of Plymouth, New Hampshire, and transmitted the following idiosyncrasy to her descendants. It is one, she observed, to which her family is unfortunately subject, and had been the source not only of great solicitude, but frequently the cause of death. If the least scratch is made on the skin of some of them, as mortal a haemorrhagy will eventually ensue as if the largest wound is inflicted. The divided parts, in some instances, have had the appearance of uniting, and have shown a kind disposition to heal; and, in others, cicatrization has almost been perfect, when, generally about a week from the injury, an hemorrhagy takes place from the whole surface of the wound, and continues several days, and is then succeeded by effusions of serous fluid; the strength and spirits of the person become rapidly prostrate; the countenance assumes a pale and ghastly appearance; the pulse loses its force, and is increased in frequency; and death from mere debility, then soon closes the scene. Dr. Rogers attended a lad, who had a slight cut on his foot, whose pulse was full and frequent in the commencement of the complaint, and whose blood seemed to be in a high state of effervescence. So assured are the members of the family of the terrible consequences of the least wound, that they will not suffer themselves to be bled on any consideration, having lost a relation by not being able to stop the discharge occasioned by this operation.
>
> The persons subject to this hemorrhagic disposition are remarkably healthy, and, when indisposed, they do not differ in their complaints, except in this particular, from their neighbors. No age is exempt, nor does anyone appear to be particularly liable to it. The situation of their residence is not favorable to scorbutic affections or disease in general. They live, like the inhabitants of the country, upon solid and nutritious food, and when arrived at manhood are athletic, of florid complexions, and extremely irascible. (Otto, 1803)

Otto observed that only males exhibited this bleeding tendency:

> It is a surprising circumstance that the males only are subject to this strange affection, and that all of them are not liable to it. Some persons, who are curious, suppose that they can distinguish the bleeders (for this is the name given to them) even in infancy . . . Although the females are exempt, they are still capable of transmitting it to their male children . . . When the cases shall become more numerous, it may perhaps be found that the female sex is not entirely exempt; but, as far as my knowledge extends, there has not been an instance of their being attacked. (Otto, 1803)

The same type of symptom was reported from another family:

> A.B. of the State of Maryland, has had six children, four of whom have died of a loss of blood from the most trifling of scratches or bruises. A

> small pebble fell on the nail of a fore-finger of the last of them, when at play, being a year or two old; in a short time, the blood issued from the end of the finger, until he bled to death. The physicians could not stop the bleeding. Two of the brothers, still living, are going in the same way, they bleed greatly upon the slightest scratch, and the father looks every day for an accident that will destroy them. Their surviving sister shows not the least disposition to that threatening disorder, although scratched and wounded. (Otto, 1803)

In vain, physicians had tried many different treatments to halt this life-threatening loss of blood. Otto reported that he had discovered that sulphate of soda was "completely curative of the hemorrhages." Its mode of operation remained unknown, but Otto argued that "a doubtful remedy is preferable to leaving the patient to his fate" (Otto, 1803).

The careful description of the clinical symptoms of the bleeding disease elicited great interest among Otto's colleagues in the medical profession. His striking observation that the females in these families were healthy themselves but could transmit the malady to some of their male offspring encouraged other physicians to investigate similar disorders in other families (Krumbhaar, 1930; McKusick, 1962).

In 1794 Elihu Smith of New York had written to Rush describing a case of fatal hemorrhage that affected his own cousin. Rush recalled the letter after learning of Otto's work, and it was eventually published in the first volume of a new medical journal, the *Philadelphia Medical Museum*. The patient had been the son of Keziah Smith, sister of Elihu Smith's father. The tragic circumstances of the boy's brief life were recounted in the case description:

> James Hawley, was born October 8, 1758 well, and a perfect child . . . Three weeks before the infant was one year old, he fell, and ruptured the fraenum of the upper lip. The lip was filleted down . . . the bleeding continued, without the least intermission, till the birthday and then stopped spontaneously . . . Several days before the child was three years old . . . the child trode [*sic*] upon a knife, and made a gash, near the little toe, about an inch in length, but scarcely skin-deep . . . The part was bound up; but bled all night. In the morning a physician was sent for, who had said that he could stop the bleeding. He attempted; but ineffectually . . . Nineteen days before the boy was to be four years old, in drawing an ax out of a log, he let it fall upon his foot; by which the little toe of the right foot was cut through the bone . . . The child died . . . five days before his anxiously-expected birthday. (Smith, 1805)

No other family members had similar bleeding tendencies. Thus the hemorrhagic disorder could occur even without antecedents in the family.

John Hay of Massachusetts later reported an extensive family of bleeders in 1813. He was able to collect genealogical information for the family which showed evidence of the character as early as 1640. The trait was reported through at least five generations. Only males were affected. Women always were physically healthy and yet produced some male children who had the bleeding disorder. Although unaffected daughters could marry and produce affected sons in the following generation, affected males never fathered affected children. Hay remarked that "children of bleeders are never subject to this disposition, but their grandsons by their daughters" are (Hay, 1813).

Another Massachusetts family of "bleeders" was described by William and Samuel Buel. In their report, normal parents had borne six children. All of the boys were afflicted and died of hemorrhages. Two daughters were healthy. One married and had normal children of both sexes. The second had six children: her two daughters were once again healthy, while all four sons were affected (Buel, 1817). As in the previous case records, the females appeared to transmit the trait, although they had no bleeding tendency themselves. Not all males born of such women were necessarily affected, but in certain instances, as in the families noted by Hay, all of the boys in two generations had received the characteristic disease from their carrier mothers.

The investigations of hemophilia by these U.S. physicians were recognized by European observers as important contributions to the medical knowledge of the era. Otto's paper was reprinted in the *Medical and Physical Journal* in England during 1808. In 1820 C. F. Nasse reviewed all the published reports on hemophilia, including these several U.S. families. He described the consistent pattern of female carriers who were clinically well and then bore affected sons. Their unaffected daughters often produced affected sons of their own in subsequent generations (Nasse, 1820). This mode of disease segregation has come to be known as Nasse's rule.

The careful analysis of the family history in these cases of hemophilia established a method for investigating the familial occurrence of other human diseases. During these same years Ennalls Martin of Maryland studied a family with "hereditary blindness" who had emigrated to the United States from France around 1650. Affected persons could be male or female. They appeared clinically well in all respects as children, but progressive blindness began by age fifteen or sixteen and was complete by age twenty-two. The first affected individual was a male who had three sighted brothers and two sighted sisters, and whose parents and relatives all were sighted. When he married and fathered eleven children, two sons were sighted, but six sons and three daughters became blind. Several of the blind sons and daughters had children of their own, some of whom were sighted and some affected. In the present generation, seventeen children had become blind. When the sighted offspring bore children, all were sighted. Martin commented that "there never has been an instance, where any of the family, who had fortunately escaped blindness, has had any blind children or that their descendants have been subject to blindness" (Martin, 1809).

The pattern of inheritance in this family was quite different from that in families afflicted with hemophilia. Both sexes were affected, and direct transmission of the character from parent to child was evident in several generations. Various physicians had tried different remedies to retard the progression of the blindness. "Frequent and copious" bleeding appeared to be most effective. In one instance, this therapy was said to have prevented total blindness until age thirty-eight.

The U.S. physicians who reported familial cases of hemophilia did not attempt to attribute a cause for the unusual bleeding tendency in their patients. Martin at least raised the question of why the vision in his cases failed at such an early age. "That there is a natural defect in the organization of the eye, communicated from parent to child is very obvious, and that the defect is in the system of absorbents of that organ is equally obvious; but why it is so, will continue to be as uncomprehensible as the mystery of creation itself" (Martin, 1809).

Martin's investigation of the family history and his attempt to discern causes for a particular disease defined the beginning, nearly two centuries ago, of human heredity applied to medical practice in the United States.

THERAPEUTICS VERSUS MEDICAL SCIENCE, 1830 TO 1860

The physicians' quest for vigorous treatments to cure their patients continued. The extensive use of bleeding as primary therapy for many ills was modified somewhat, as experiments in France by Pierre Louis and others demonstrated that massive bleeding did little for patients with pneumonia and other inflammatory diseases. U.S. physicians studying in Paris at this time agreed with these observations and carried them back to the United States. Thus the generation of U.S. physicians trained in France became less inclined to use bleeding routinely for all kinds of illness.

Most physicians combined more moderate bleeding with ever-increasing doses of medications, hoping to obtain the benefits from both types of therapy. Tartar emetic (antimony and potassium tartrate) was believed to have the "remarkable property of subduing inflammatory actions in the internal organs of the body." Calomel (mercurous chloride) was used as a "bilious purgative"; it purportedly helped cleanse the patient of hepatic, pancreatic, renal, intestinal, salivary, and cutaneous secretions. Arsenic preparations were another important medication, widely used for skin diseases, dyspepsia, backache, epilepsy, cancer, scrofula, dysmenorrhea, and rheumatism. The practical role of the physician of the day was to bleed, purge, blister, or puke the patient until he or she either died or persevered long enough to recover from the combination of the disease and the therapy rendered (Haller, 1981).

The education of physicians in the United States deteriorated after 1820, when the apprenticeship system was replaced by brief periods of training in proprietary medical colleges. With few or no standards, these were organized to train large numbers of students and to make money for the faculty. The democratic society encouraged anyone to become a physician. State licensing requirements, which had been established earlier in the century, were gradually abolished. By 1860, with no effective licensing of physicians in any of the states, anyone could practice medicine regardless of education, experience, or moral character (Haller, 1981).

This cheapening of medical education had reduced the profession to the lowest level. Those U.S. graduates who could afford the European tour continued their education there, and Paris evolved as the center for postgraduate medical education of young U.S. physicians. They observed patients in clinics and hospitals and performed autopsies and biopsies to study cellular histology as it correlated with clinical symptoms of disease. The leaders of French medicine encouraged them to observe carefully and then to apply statistics in analyzing their clinical data (Haller, 1981; Harvey, 1981; Ludmerer, 1985).

Modern concepts of the causality of disease emerged from this work in France. The application of physical diagnostic skills such as percussion and auscultation and the careful observation of the natural history of disease progression permitted the grouping of abnormal findings to define a disease state. The use of pathological anatomy—

whether biopsy during life or autopsy after death—encouraged early attempts to define changes within the body's anatomy which produced the symptoms observed via physical examination of the living patient (Cassell, 1979; Harvey, 1981). The major purpose of this clinical-pathological correlation was to better understand the processes of different disease states. Less attention was devoted to curing the diseases that had been elucidated.

Young U.S. physicians who trained in Paris returned home eager to apply these new ideas to medical education and practice. But they received a generally cool reception from their colleagues, who distrusted the French interest in medical science rather than therapeutics. For example, one prominent U.S. student in Paris at this time was James Jackson, Jr., whose father was physician-in-chief at the Massachussets General Hospital. The younger Jackson had studied with Pierre Louis, a proponent of careful clinical observation of disease symptoms. Louis gathered as much data about each patient as possible and then correlated symptoms with anatomical lesions observed at autopsy. His work early in the nineteenth century helped transform medicine into a modern science (L. King, 1991).

Louis wrote to James Jackson and urged that his son be allowed to spend time doing research rather than be pushed immediately into medical practice. The senior Jackson would hear none of it, "because," he wrote to Louis, "in this country, his course would have been so singular, as in a measure to separate him from other men. We are a business-doing people. We are new. We have as it were just landed on these uncultivated shores: there is a vast deal to be done, and he who will not be doing must be set down as a drone" (Bowers, 1976).

As the primary mission of physicians was to cure patients, anything that diverted their attention from this goal was viewed with suspicion. Many practicing physicians feared that devotion to "science" as exemplified by the European models would ultimately corrupt the character of the U.S. professional in the healing art: "Physicians should never resemble those philosophers who pursue medicine in the abstract, as all pursuits of practicing physicians should have a practical tendency" (Warner, 1985).

Medical science at this time did not appear to offer any improvement over existing therapies for the diseases encountered in the United States. Further, its pursuit could lead the physician away from caring for patients. Because science thus appeared both unnecessary and potentially destructive of current medical practice, little enthusiasm was generated for these new ideas among most U.S. practitioners of the day.

Another reason why U.S. physicians distrusted the impact of modern science on medical practice was their lack of formal training in this area. The medical school curriculum was eminently practical and encouraged little exploration of the basic sciences. Some physicians did find intellectual stimulation in the study of natural history—botany, geology, zoology, and agriculture—homegrown sciences that encouraged them to observe closely and to organize their thinking processes. J. K. Mitchell of Jefferson Medical College in Philadelphia suggested to medical students that they improve their minds by seeking more than the didactic education offered in the classroom. He argued that interest in natural history did not hinder professional medical activity but broadened practitioners' minds as they sought to understand the whole of nature better. Such investigation also enlivened the mind and encouraged the physician to observe patients in a more detailed fashion than less knowledgeable practitioners, which, Mitchell believed, better served the patient (Haller, 1981). In the absence of formal science education, this local experience may well have fostered logical thinking and may ultimately have made U.S. physicians somewhat more receptive to the advances in science which were evolving in France and Germany.

HEREDITARY COLOR BLINDNESS

Pliny Earle was a practicing physician of this era who reflects the congruence of the Americans' naturalistic experience with detailed European medical training. Published in 1845, his analysis of color blindness represented a significant step in the evolution of U.S. medical science in general and in the understanding of human heredity in particular.

Earle was a native of Massachusetts, born in 1809. His family owned a large farm, and his father designed machinery for the cotton-weaving industry. The family virtues of patience and perseverance would ultimately be reflected in Earle's life. He was first tutored at home, then completed his studies in 1829 at the Friends School in Rhode Island and was hired there as a teacher. He became interested in biology and led students on collecting trips to the seashore and local forests, encouraging the systematic cataloging of everything found on these excursions. At the same time he began the private study of medicine with Usher Parsons, brother-in-law of Oliver Wendell Holmes.

In 1835 he enrolled at the University of Pennsylvania Medical School, where about four hundred students attended the lecture series that year. He quickly discovered that medical students were often held in disdain by Philadelphians. Not only were students not inclined to study hard, but they were often cruel and lawless—stabbing or shooting at each other or hapless citizens of the town. Earle completed the two-year required lecture course and departed for further training in Europe in 1837.

Most of the next year he spent in Paris attending lectures by Pierre Louis and other physicians who specialized in mental and neurologic disorders. He toured the city's insane asylums and every day for months made hospital rounds to examine patients. Keeping careful notes of all he observed, he was impressed with the small doses of medicines used in France compared to the heroic treatments then in vogue in the United States.

In 1840 Earle returned to Philadelphia and began his career caring for mentally ill people. For the next forty-four years he superintended several mental institutions and sought to humanize medical care for the insane and mentally ill. He abolished bleeding and heavy restraints for insane patients and performed pharmaceutical trials on himself and his patients to test the alleged benefits of medications used for mental illness. To encourage self-help for the inmates, he established schools and workshops within the institutions. Through the years he kept careful statistics on his patients and attempted to draw conclusions on the efficacy of treatment options. Toward the end of his career, in 1884 he was elected president of the Association of Medical Superintendents of Institutions for the Insane (Sanborn, 1898).

Earle's interest in color blindness developed after he learned that many individuals within his own family carried the trait. His approach to hereditary analysis was far in advance of anything else reported in the U.S. medical literature of that era.

In the manuscript upon which he based his final report (Earle, 1845), reproduced in the biography by Sanborn (1898), Earle includes extensive material not incorporated in the published paper. He first reviewed seventeen cases of color blindness from the European literature. Males were noted to lack the ability to distinguish certain colors. Some individuals could recognize only black or white. Others could identify one or two colors of the spectrum, but not all. The disorder appeared to be a familial trait that was transmitted through unaffected females. One family had one affected son whose father, mother, and four sisters had normal vision for color. The mother's father, who was color-blind, had one brother who exhibited the defect and another brother and a sister with normal vision for color.

The inability to distinguish red and green was evident in many individuals from Earle's family: "A child ran into the room of his grandmother, where scraps of variously colored paper were lying on the floor. 'Oh, ' exclaimed he, in childish joy, 'here's some red paper, ' and immediately collected all the pieces of green. When he became old enough to wield a pencil, he manifested some skill in drawing, but the yellow bears, and the black birds of paradise, and the green men and ladies with green cheeks, red eyes, and blue hair, that were brought into existence by his truly original genius would have . . . made Titian and Raphael believe they had mistaken their calling" (Sanborn, 1898).

Earle was able to collect information on five generations to produce the most extensive family history of color blindness that had been published to date:

> Nothing is known in regard to the first of five generations in regard to color blindness. In the second of a family consisting of seven brothers and eight sisters, three of the brothers, one of whom was my grandfather, had the defect in question. In the third generation, of three brothers and six sisters, there was no trace of the defect. In the fourth generation, the first family is composed of five brothers and four sisters, and two brothers have the defect. In the second, there was but one child, a daughter, whose vision was normal. In the third, there are seven brothers, of whom four have the defect; in the fifth, seven sisters and three brothers, in all of whom the vision is perfect in this respect; in the sixth, four brothers and five sisters, of whom two brothers and two sisters have the defect; in the seventh, two brothers and three sisters, both of the former having the defect. In the eighth, there was no issue, and in the ninth, there are two sisters, both of them able to appreciate colors. (Sanborn, 1898)

Earle summarized these findings in a published article in a pedigree chart, a family tree in which circles represented females and squares represented males. Affected individuals were noted by blackening the symbol. Earle drew several conclusions from his analysis of the family data. First, color blindness affected males much more frequently than females. Of twenty cases in the family, all but two were in males. Of all the lines in which the defect was present, there were thirty-two male offspring, and eighteen (or nine-sixteenths) were afflicted. Of twenty-nine females in the same lines, only two (or one-fifteenth) exhibited the disorder. Another peculiarity about this disorder was its "overleaping of one generation" or more: in the third generation, no individuals were affected, and yet in one branch of the fifth generation, two males were affected, while neither the parents nor the grandparents in between had the defect. Thus one or two generations could transmit the character and yet remain unaffected (Earle, 1845).

The use of the pedigree to detail the family history was a first in U.S. medical literature. Although Earle provided no reference for his knowledge of pedigree construction, he may have encountered it during the time he studied in France. Family studies of polydactyly by Pierre Louis Moreau de Maupertuis and R. A. F. de Réaumur in the previous century had been summarized by the use of such family trees (Réaumur, 1751; Maupertuis, 1768; Glass, 1968). Earle's use of statistics to organize his thinking, a carryover from his days as a schoolteacher, was also undoubtedly fostered by the analytical approach to medicine which he encountered during his experiences in Paris.

In a departure from the descriptive nature of previous reports on human heredity, Earle sought to provide an explanation for the defective vision that he had observed. He considered theories for color vision proposed by other physicians and reviewed the organic defects that could occur in either the visual apparatus or the organ of visual perception, the brain. The eyes of color-blind individuals appeared to be formed normally in all other respects. He concluded that there was probably a cortical defect in the brain which prevented the accurate perception of color by these individuals (Earle, 1845; Sanborn, 1898).

At this point in the analysis of heredity, the science of the day had reached its limit. No further explanation for the transmission of this apparent hereditary defect could be given. But some general rules for the action of heredity were being defined within the general practice of medicine.

HEREDITY AND MEDICINE

It was obvious to physicians early in the nineteenth century that certain traits were hereditary: they ran in the family. Heredity was usually summarized as "like begets like" (Rosenberg, 1967). In 1847, Harris felt confident enough to remark that "laws of heredity are settled and acknowledged" (Harris, 1847). Two basic patterns for the action of heredity were recognized. The first was the notion of hereditary disease, which assumed that the fetus had received "morbific seeds" at the commencement of its existence and was born with an abnormal characteristic. Generally, this was believed to be a rare occurrence. The second was the idea of hereditary predisposition, felt to be much more common. In this instance, a child who appeared to be healthy at birth had inherited "some hidden weakness of certain organs by which these are prone to take a diseased action" (Brown, 1845). The malfunction of specific organs was the basic definition for disease at this time, and most physicians agreed that this type of disease was not inevitable but the result of an exciting cause that produced symptoms of disease in individuals with the predisposition. Some people with the constitutional tendency never developed disease symptoms because they never encountered the specific trigger during their lifetime. But they were certainly liable to transmit the hereditary defect to their offspring, "who may break out with the old disease of the grandparents" (Lewis, 1860–63).

The first humans were believed to have been healthy, but the accumulation of vices over generations had weakened the stock and rendered modern humans incapable of propagating only healthy offspring (Hayes, 1841–43). There was general agreement that characteristics acquired in one generation could be transmitted to the next (Ogier, 1848; Pallen, 1856), but the management of these hereditary predispositions was not viewed as hopeless by many physicians. Inherited qualities could be attenuated by avoiding inciting causes when family members were known to be at risk. Good maternal care after birth was also believed to be important for the establishment of good health during later years (Ogier, 1848).

As studies reported before the Civil War show, both internal and external human features appeared to have a hereditary basis (table 1.1). To decrease the likelihood of such predispositions, physicians generally recommended the avoidance of marriage in families with any of these "hereditary taints" (Brown, 1845; Lewis, 1860–63).

Physicians in the United States were also becoming aware of the developments in embryology which were relevant to their discussions of heredity. Ogier remarked that current opinion suggested that the ovum produced the fetus, while the sperm merely excited or stimulated the process of development, although he did not agree with this thesis. He observed that in fact both maternal and paternal traits appeared in the human offspring, implying some contribution from both parental lines (Ogier, 1848). Other physicians reported that the biologists really could not explain how the single-celled egg (composed of "albumen and oil globules") could develop into the complex called a human (Pallen, 1856).

Table 1.1. Hereditary Predispositions
Reported, 1840–1860

Angina	Facial features
Apoplexy	Gout
Asthma	Hemorrhage
Blindness	Idiocy
Cancer	Insanity
Cataracts	Myopia
Chorea	Nervous disorders
Complexion	Polydactyly
Consumption	Scrofula
Deafness	Stature
Epilepsy	Syphilis

Sources: Data from Brown, 1845; Harris, 1847; Ogier, 1848; Bowen, 1860–62; Lewis, 1860–63.

CONCLUSION

The understanding of heredity by U.S. physicians changed dramatically during the first half of the nineteenth century. The working definition of heredity (Like begets like) had been demonstrated by familial disorders such as hemophilia and blindness. The distinction between hereditary disease, which was uncommon but present at a young age, and hereditary predisposition, which was much more frequent and acted later in life, was established by 1860. To explain the cause of widely diverse human afflictions, physicians drew on the notion of predisposition or diathesis that required an exciting trigger to produce clinical symptoms of disease.

The use of the pedigree, review of previously published cases, and attempts to define the anatomical defect that produced the altered function of specific organs signaled the development of a sophisticated and more scientific understanding of heredity by some U.S. physicians. Heredity was clearly associated with embryologic development as well, and no study of heredity was undertaken without an attempt to understand how such transmitted elements could have altered normal embryonic development. The abnormal "soil" then could become the source of symptoms of malfunction or disease, if the individual was unfortunate enough to encounter the necessary trigger. Whether because of scientific training in Europe or the logical thinking encouraged by domestic natural history studies, U.S. physicians had begun to demonstrate their ability to appreciate the potential of science for medical practice.

TWO

THE FORMATIVE YEARS
1860 to 1900

The late nineteenth century was a period of momentous change in U.S. medicine. The elaboration of the germ theory and basic research into disease mechanisms provided more rational approaches to the diagnosis and treatment of many disorders. These changes in medicine paralleled developments in biology. Over time, the application of research findings from biology profoundly altered the nature of daily medical practice.

By 1860 German universities had organized divisions of medicine within the general realm of the biologic sciences. Original laboratory research was encouraged as a key to understanding the basic causes of human disease. Such scientific data were no longer viewed by physicians as a mere curiosity. Instead, they were regarded as the basis for careful diagnosis and eventual successful treatment of disease. Young U.S. physicians studying in Germany at this time were caught up in the excitement of medical discovery. They became convinced that "medical phenomena were governed by chemical and physical laws." Controlled laboratory experiments, not merely clinical observation of the patient, shed light on pathology. Certainly much of the new scientific information may have seemed esoteric and rather distant from any immediate application to the care of specific patients. But by the end of the century both medical educators and practitioners had become convinced that experimental medicine could aid in the chief role of the physician, which was always therapeutic—to treat human disease and ease patients' suffering (Bonner, 1963; Ludmerer, 1985; Warner, 1986).

Many physicians of the era also recognized that the vigorous therapy that had been used to dramatically alter patients' clinical symptoms often had no effect on their eventual recovery. Because medicine was not yet an exact science, some argued, it might be better to let nature take its course, rather than to rely on the sometimes hazardous ministrations of physicians (Rothstein, 1972). A well-known essay by Jacob Bigelow of the Harvard Medical School defined "self-limited diseases" as those that tended to run their course, either to recovery or to death, regardless of intervention by the medical practitioner. He proposed that physicians "should not allow [the patient] to be tormented with useless and annoying applications, in a disease of settled destiny. It should be remembered that all cases are susceptible of errors of commission as well as omission, and that by an excessive application of means of art, we may frustrate the intentions of nature, when they are salutary, or embitter the approach of death when it is inevitable" (Bigelow, 1854).

This style of medical practice was not popular with the public or with most physicians. In a speech before the Massachusetts Medical Society Wyman labeled Bigelow a therapeutic "extremist" and "nihilist." He argued that the term "self-limited disease" really meant that no therapy was available. Malaria, once thought inevitable, could now be successfully treated with quinine (Wyman, 1863). Certainly the effort to do something for the patient engendered hope, which always aided recovery from illness (Rothstein, 1972).

DEVELOPMENTS IN HEREDITY

The traditional history of genetics has viewed the modern understanding of heredity as the result of a series of logically connected discoveries by experimental biologists after 1850 (reviewed by Sturtevant, 1965; Carlson, 1966; and Stubbe, 1972). Mayr (1982) and Bowler (1989) more recently emphasized the importance of major changes in conceptual thinking—paradigm shifts—before these new discoveries made sense. Heredity in the early part of the nineteenth century was not a distinct area of scientific investigation; it was viewed as a part of the general area of growth and reproduction. Offspring were usually like their parents. Hereditary factors were passed on to progeny and could be altered by injury, use, or disease. The inheritance of acquired characteristics was accepted fact.

The advent of experimental embryology in Germany around 1850 produced a series of discoveries that eventually convinced many biologists that the heredity of specific characteristics was separate from development, and that development itself was controlled by factors within the fertilized ovum. At the same time all of biology moved from being a descriptive science to an experimental one. For an observation to count as fact, it had to be testable.

Observations in Europe suggested that hereditary factors from both parents affected the development of their offspring. The discovery of egg and sperm made it likely that these cells were the vehicles that somehow initiated the process of embryonic development. Pringsheim observed in 1856 that one sperm entered one ovum to produce the fertilized egg or zygote in algae (1856). Later studies by Fol (1877) and Hertwig (1884a) confirmed these observations in other species. The egg and sperm cell nuclei were seen to fuse into a single nucleus within the zygote. Subsequent division of the zygote and its nucleus resulted in the cells that eventually formed the embryonic organism. Microscopic studies of the cell nucleus revealed colored strands, which were named chromosomes by Waldeyer (1888). Flemming observed that a longitudinal split of the chromosomes occurred before cell division (1882). Roux (1883) and Van Beneden (1883) subsequently demonstrated that equal numbers of maternal and paternal chromosomes appeared in the two daughter cells after mitosis. There was no general agreement, however, on the relevance of these observations for the development of the embryo. Many biologists viewed development as a process of tissue and environmental interaction which eventually resulted in a new organism. There was much less agreement that factors within the cell nucleus were the primary determinants of embryonic development.

During the 1880s another series of observations was used to argue that the chromosomes within the cell nucleus carried the hereditary factors from one generation to the next, and that these factors were indeed responsible for guiding the process of embryogenesis. Nageli (1884), Weismann (1891–92), and Hertwig (1884b) all provided evidence that the nuclear material was the bearer of the hereditary substance.

At the same time, a major change in the understanding of transmission from one generation to another was taking place. The English biologists Spencer (1864) and Darwin (1868) proposed that a blending of factors from both parents constituted heredity. External factors could alter the composition of these units and allow for the inheritance of acquired characteristics from the parents. Darwin's cousin, Francis Galton, who attempted to verify these theories by experimental breeding of animals, could find no evidence for the transmission of acquired features (1871). He concluded that packets of hereditary material were not altered before transmitting hereditary information from parent to children.

The concept of "hard" heredity was further elaborated by experimental work on animals toward the end of the century. De Vries in Holland proposed that material entities called pangens resided in the cell nucleus and functioned as bearers of heredity from one generation to the next. Each pangen controlled a particular trait. New characters could arise spontaneously by changes within the hereditary units themselves (De Vries, 1889). Weismann in Germany believed that heredity involved the transmission from one generation to the next of factors with a definite chemical constitution. He argued that these "germplasm" factors resided in the chromosomes of the cell nucleus. Fertilization involved the union of equal numbers of factors from both parents. His experimental studies found no evidence for the heredity of acquired features. He was also convinced that variation in heredity could result only from changes within the germplasm itself (Weismann, 1892).

HEREDITY, PREDISPOSITION, AND MEDICINE, 1860 TO 1880

Before any of these discoveries would be applied by U.S. practitioners, however, their thinking would have to change. Clinical experience suggested to physicians of the post–Civil War era that certain human traits were hereditary: they ran in the family. Normal features (such as physiognomy, hair color, and stature) and disease states (such as epilepsy and cancer) both appeared common to certain families. These observations on heredity and human characteristics were discussed in many articles in the U.S. medical literature during these decades.

It was believed that much human suffering was the result of abnormal heredity. There was an increasing tendency to ascribe both physical and mental defects to heredity. If physicians observed an abnormal trait in a patient and discovered that it had occurred in a relative, they usually assumed that this showed that hereditary factors had contributed to its development (Bowen, 1860–62; Carrington, 1868–69).

Although both physical and mental qualities appeared to be transmitted from parent to child, children were never exact copies of their parents. Two conflicting forces were evidently at work in this process of heredity. One was conservative, transmitting the general features of the species from one generation to the next. The other permitted changes to occur. Those individual features of body and mind that made each person unique were also the result of the mix of characters from each of the parents. The tension between these two features of heredity resulted in both the uniformity and the diversity evident among human individuals (Ray, 1862).

Much human disease seemed to have an hereditary basis. Malformations of the organs could be transmitted, which resulted in the abnormal functioning defined as disease. Cataract, color blindness, hare lip, and extra digits all were transmitted in some families from one generation to the next. Scrofula (tuberculosis of the lymphatic glands), gout (inflammation of the joints), syphilis, and apoplexy (stroke or cerebrovascular accident) frequently recurred in one generation after another (Ray, 1862).

A large number of disorders appeared to be inherited as predispositions or diatheses. Symptoms of disease were not inevitable, since the predisposition was always viewed only as the potential for disease. An exciting trigger appeared to be necessary for the development of the disease itself (Allen, 1869; Iles, 1878). The notion of predisposition or diathesis involved the inheritance of general constitutional weakness, broadly defined.* Asthma and cancer were common hereditary diatheses. Nervous and mental diseases were also frequently present in several generations of "badly tainted families," although specific symptoms in one generation were often modified in offspring. Mental retardation or habitual intoxication in a parent, for example, might be followed by hysteria or insanity in the children (Ray, 1862; Dupuy, 1877). Because the central nervous system was so highly organized, it was believed to be particularly vulnerable to alterations of its normally healthy state brought about by defective heredity. "In general, one inherits a family proclivity to nervous disorders, in one case idiocy, in another mania, in another convulsion" (Rogers, 1869).

In other cases, there seemed to be no specific hereditary taint, but the generally defective offspring appeared to be susceptible to all kinds of disease. The children in such families were said to have "peculiar constitutional characters" that resulted in a low grade of vitality. There was not "any special hereditary disease, " but the children did not respond to sound medical care and were easily carried away by "scrofula, rheumatism, syphilis, general anemia, and sexual exhaustion" (Peters, 1879).

Therapy for those individuals predisposed to hereditary disease or general weakness involved avoiding the exciting triggers for disease whenever possible. Careful living was often able to prevent or at least ameliorate the development of clinical symptoms (Iles, 1878). Proper exercise, good food, and clean air all were helpful. One physician suggested that cod liver oil could act as a revitalizer for these predisposed individuals if it was used before symptoms of disease actually appeared. Although its mechanism of action was "somewhat myste-

---

*For review, see Rosenberg, 1967, 1974.

rious, " cod liver oil was believed to "revolutionize the whole body, building up every cell." It reorganized the vital forces and made the person into "another being" (Peters, 1879).

The relative importance of the parents in the transmission of these predispositions was a controversial question. Heredity was believed to be just one aspect of human reproduction and embryonic development in general. The theories of heredity advocated by Spencer and Darwin were used to describe how factors in the egg and sperm united to form the fertilized egg or zygote and eventually guided the structural development of the embryonic organism (Carrington, 1868–69). Other physicians believed that the sperm did not unite with the egg but merely activated it, initiating the course of embryogenesis (Stockton-Hough, 1875). The outcome of heredity was always viewed as a dynamic process ranging over time from the fertilization of the egg to birth and beyond for several years. Environmental factors during gestation, such as maternal anxiety, grief, or illness, all conspired to alter the expression of the hereditary units in the fertilized egg (Iles, 1878).

The mother and father appeared to have different roles in the hereditary process. Stockton-Hough observed that the mother contributed the most to the eventual development of the children of both sexes. The determination of sex was related to the age of the ovum when it was activated by the sperm. Females developed from ova fertilized shortly after ovulation, while males were the result of slightly older ova. Thereafter, the mother impressed upon the developing fetuses physical and moral peculiarities, constitutional tendencies, and hereditary diseases. The general resemblance of these traits to the mother was believed to be much more marked than that to the father. It was noted that the ova had existed in the female from early in her life. Impressions or acquired alterations that accumulated over the years were likely to occur in the egg and then could be transmitted to the children. Sperm, on the other hand, which did not develop until puberty, had a much shorter period in which to acquire new characters. Therefore, the female was much more likely to transmit acquired impressions than was the male. The molding role of the mother then continued throughout gestation and thereafter, as she had exclusive control during the formative early years of the child's life (Stockton-Hough, 1875).

Physicians of the day, like their biologist counterparts, viewed heredity as a process, not as an event in time. The working definition of hereditary descent continued to be "Like begets like" (Allen, 1869), as many examples of reports from the clinical literature reflect. One study, for instance, reported that the grandfather, parent, and three children in one family all had dwarfed upper incisor teeth (McQuillen, 1870). Another noted that consumption (pulmonary tuberculosis) frequently recurred in successive generations of specific families; when present at birth, this was felt to provide conclusive evidence that the disorder was hereditary in nature (Cleveland, 1877). Familial cases of fibroma molluscum, benign tumors of the peripheral nerves, also were reported at this time. The recurrence of the character in parent and child afforded "pretty conclusive evidence that heredity may play a part in the etiology of the affection" (Atkinson, 1875).

An early attempt to classify patterns of human inheritance was made by the French physician Theodore Ribot, whose work appeared in English translation in 1875 and defined four basic modes of inheritance:

1. Direct: from parent to child;
2. Indirect or reversional: an atavistic reappearance of a trait seen in a previous relative, such as a grandparent;
3. Collateral: the appearance of a trait found in another line of the family, such as an uncle or cousin;
4. Heredity of influence: the appearance of a trait in the offspring of a subsequent marriage which was similar to that of the mother's first husband.

While this paradigm did not outline mechanisms of heredity, it did prove useful for U.S. physicians as they attempted to organize their thinking about human inheritance over the next thirty-five years (Dolan, 1888–89; Walker, 1897–98).

Examples of direct heredity were reported for both normal structural variants and disease states. The presence of extra digits (polydactyly) was observed in four successive generations (Brown, 1873–74; Tomlinson, 1879–80). The course of the radial artery was aberrant in three generations of another family, a "remarkable example of hereditary transmission of an anatomical peculiarity" (Schneck, 1879). Neurologic disease was often seen in both parent and child. A particularly striking case involved a family in which the mother and nine children were afflicted with epilepsy, and all of the children died by age thirteen months (Gray, 1879). Muscular atrophy was evident in brothers and sisters of another large family. Some of their children were also afflicted, while the offspring of unaffected persons were always normal (Osler, 1880).

A special type of indirect heredity involved unaffected mothers transmitting a "constitutional tendency" to some of their sons. Hemophilia was observed to follow this pattern in many families (Whittaker, 1880), although female bleeders did occasionally appear. Both sons and daughters were affected in one such family over five generations (Morrison, 1881). The same pattern of inheritance was noted in families with pseudohypertrophic muscular paralysis (Duchenne disease). Boys often developed symptoms of muscle weakness at age five or six. The character was transmitted only via the mother's side of the family (Poore, 1875; Steele, 1879; Rowland, 1881).

Color blindness often followed a similar pattern (DeFontenay, 1881), but affected females were reported (Becker, 1880). When both sides of the family demonstrated color blindness, the likelihood of an affected female appeared to be increased (Jeffries, 1879). Different ethnic groups also had variable frequencies of color-blind children. One report documented the disorder in 1.6 percent of black males, while another study found that 2.5 percent of white males and 1.8 percent of Native American males had the disorder. The expression of this hereditary predisposition was believed to be ameliorated by the "earlier education of color in infancy" received by the Native American children" (Fox, 1882).

Examples of the indirect and collateral patterns of heredity defined by Ribot came to be viewed by physicians as indicative of familial traits that were not in fact hereditary, so that a more restrictive definition of what constituted a hereditary character gradually developed in the last quarter of the nineteenth century.

HEREDITARY CHOREA

An unusual degenerative disease of the central nervous system was recognized during this period as the paradigm of an hereditary disorder that was transmitted directly from parent to child. The classic description of "hereditary chorea" was published in 1872 by a young Ohio physician named George Huntington. Toward the end of the century, William Osler drew attention to the importance of this report with the remark that "in the history of medicine, there are few instances in which a disease has been more accurately, more graphically or more briefly described" (Osler, 1894b).

Huntington came from a family of physicians on Long Island, New York. His grandfather settled there in 1797 and observed a strange malady in certain families residing in the community. Young George was introduced to it at the age of eight.

In riding with my father on his professional rounds, I saw my first cases of "that disorder." I recall it as vividly as though it had occurred yesterday. It had a most enduring impression on my young mind, an impression every detail of which I recall today, an impression which was the very first impulse in my choosing chorea as my virgin contribution to medical lore . . . We suddenly came upon two women, mother and daughter, both bowing, twisting and grimacing. I stared in wonderment, almost in fear. What could it mean? . . . My medical education had its inception. From this point, my interest in the disease never wholly closed. (Huntington, 1910)

Huntington was educated by local tutors and began to learn medicine from his father as an apprentice. At twenty-one, he graduated from the College of Physicians and Surgeons in New York City (Kaufman, Galishoff, and Savitt, 1984) and returned to Long Island to assist his father in medical practice. Building on notes his father and grandfather had accumulated on patients with chorea over the years, he prepared a preliminary draft of a report on this unusual malady during 1871. Later that year he moved to Pomeroy, Ohio, upon the advice of a relative and sought to establish his own practice there. His paper on chorea was read before the Meigs and Mason Academy of Medicine at Middleport, Ohio, in February 1872 and was published later that year (Huntington, 1872).

In it, Huntington described the essential features of the disorder that had captured his attention from boyhood:

The hereditary chorea, as I shall call it, is confined to certain and fortunately few families, and has been transmitted to them, an heirloom from generations away back in the dim past. It is spoken of by those in whose veins the seeds of the disease are known to exist, with a kind of horror, and not at all alluded to except through dire necessity, when it is mentioned as "that disorder." It is attended generally by all the symptoms of common chorea, only in an aggravated degree, hardly ever manifesting itself until adult or middle life, and then coming on gradually but surely, increasing by degrees, and often occupying years in its development, until the hapless sufferer is but a quivering wreck of his former self.

There are three marked peculiarities in this disease. 1. Its hereditary nature. 2. The tendency to insanity and suicide. 3. Its manifesting itself as a grave disease only in adult life.

1. Of its hereditary nature. When either or both the parents have shown manifestations of the disease, and more especially when the manifestations have been of a serious nature, one or more of the offspring almost invariably suffers from the disease if they live to adult age. But if by any chance, these children go through life without it, the thread is broken and the grandchildren and great grandchildren of the original shakers may rest assured that they are free of the disease . . . It never skips a genera-

tion to manifest itself in another; once having yielded its claims, it never regains them.

2. The tendency to insanity, and sometimes that form of insanity which leads to suicide, is marked.

3. Its third peculiarity is its coming on, at least as a grave disease, only in adult life. I do not know of a single case that has shown any marked signs of chorea before the age of thirty or forty years, while those who pass the fortieth year without symptoms of the disease are seldom attacked. It begins as an ordinary chorea might begin, by the irregular and spasmodic action of certain muscles, as of the face, arms, etc. These movements gradually increase, when muscles heretofore unaffected take on the spasmodic action, until every muscle of the body becomes affected (excepting the involuntary ones), and the poor patient presents a spectacle which is anything but pleasing to witness . . . No treatment seems to be of any avail. (Huntington, 1872)

The striking clinical features of the disease were unmistakable. Earlier reports had briefly described the malady. Another New York physician, C. O. Waters, reported these spasmodic muscular symptoms in affected individuals in 1842. He called it "markedly hereditary." Waters noted that he had never known it to occur in a person who did not have parents and grandparents who also were subject to this "distressing malady" (Waters, 1842). Irving Lyon of Bellevue Hospital in New York City also reported three affected families in 1863. Again the disorder appeared to be transmitted directly from parent to child. In one family, five successive generations were affected. Lyon argued that the existence of the disorder in so many generations was difficult to explain except on an hereditary basis (Lyon, 1863).

The report by Huntington reflected his own background and that of many other physicians of the age. He grew up in a rural setting and learned to be a careful observer of nature. He knew all the birds, trees, and flowers in the area. He taught himself to sketch and paint and produced realistic representations of local geese, ducks, and pheasants. His description of hereditary chorea reflects these skills, for he accurately reported all that he saw in succinct language. But his report is not scientific. He knew nothing of the biologic theories of Spencer or Darwin, which could have provided a mechanism to explain the transmission of this character from parent to child. He did not attempt to investigate changes within the structure of the central nervous system which resulted in the clinical symptoms of the disease. He never appears to have considered that a postmortem examination of one of the local victims might provide useful insight into the underlying cause of these distressing neurologic symptoms. His medicine was studious and meticulous, but not incisive. He was never moved by the new scientific thrust of medicine to learn more about the pathologic mechanisms of disease. Only later would physicians become convinced of the importance of this scientific attitude—that it could eventually improve their ability to care for the illnesses of their patients.

SCIENCE, EDUCATION, AND THE U.S. PHYSICIAN

Changes in medical education in the United States paralleled changes in college and graduate education after the Civil War. Students were encouraged to become not merely passive recipients of lecture material but active participants in the creative process of scientific research that came to characterize the best type of education in general (Servos, 1986). After 1870, colleges such as Williams, Yale, and Princeton offered science laboratory training to students. Graduate programs in the sciences were also established at leading schools such as Yale, Harvard, Pennsylvania, and Johns Hopkins to offer research training leading to the Ph.D. degree. This process eventually reduced the necessity for U.S. students to travel to Europe to obtain a good science education (Bruce, 1987).

Premedical Education

The preceptor training that had provided most of the applicants for medical school admission gradually disappeared by 1840. After that time, almost any man could enter medical school, pay tuition, and graduate with little scientific training, a situation that led the American Medical Association, from its inception in 1847, to advocate improved preparation for the study of medicine. The group recommended that each student obtain "a good English education, a knowledge of natural philosophy and the elementary mathematical sciences, including geometry and algebra, and such an acquaintance at least with the Latin and Greek languages as will enable [him] to read and write prescriptions" (Numbers, 1988). If these recommendations had been implemented, the vast majority of medical students in the United States before the Civil War would not have qualified. There were essentially no state licensing requirements for physicians, and whenever a medical school attempted to implement stricter requirements for graduation, its students merely transferred to another school with a less stringent curriculum (Kaufman, 1960; Numbers, 1988). No particular advantage accrued to those with extensive medical educations.

The physician of this era was not a scientist but a practical tradesperson (Rothstein, 1987). Several universities offered premedical courses in the sciences, but these were unpopular with students. For example, President Daniel Gilman of Johns Hopkins University believed that the quality of U.S. physicians would be improved if they received better education in the natural sciences. He appointed H. N. Martin as professor of biology in 1878 for the express purpose of providing the scientific "instruction antecedent to the professional study of medicine." This undergraduate curriculum offered work in histology, biology, comparative anatomy and physiology, chemistry, physics, and foreign languages (Harvey, 1981; Pauly, 1984). But the medical school at Johns Hopkins was unable to open until 1890. Not only were there no medical faculty on campus to stimulate research projects that could involve the undergraduate students, but few students appeared inclined to undergo such rigorous preparation, because it seemed to be "an unnecessary luxury for the prospective medical student" (Harvey, 1981).

Another early effort evolved at the Yale Sheffield School, organized in 1869 "to teach the principles of science, the laws of its application, the right methods of research, the exact methods of computation, analysis and observation and a fundamental reverence for truth" (Bruce, 1987). R. H. Chittenden, the first director, organized a physiological chemistry laboratory. His primary aim was to develop a program designed to train students in the methods of scientific research.

One of the first students in the new school was T. M. Prudden, who entered after studying Greek, Latin, and mathematics at a private academy in Massachusetts. During his first year at Yale, he studied French, German, English, mathematics, chemistry, and physics. At the end of that year, Prudden and another student, Thomas Russell, decided to pursue the study of medicine. No premedical curriculum existed, but the faculty organized courses in botany, zoology, embryology and "the laws of hereditary descent" (Harvey, 1981).

Both Prudden and Russell worked closely with faculty members, acting more as research assistants than as typical college students. Prudden worked in the chemistry laboratory and also made field observations on the biology of the fiddler crab. He graduated from Yale in 1872 and spent the summer collecting plant and animal specimens in Maine. During the following academic year he substituted for the chemistry professor, giving lectures and supervising the laboratory exercises. At the same time, he attended lectures at both the Yale Medical School and the College of Physicians and Surgeons in New York City, obtaining the Yale M.D. in 1875.

Prudden studied histology with Frances Delafield in New York and became convinced of the importance of laboratory research in the attempt to understand processes of disease. He spent a year as intern at the New Haven Hospital, where he organized a clinical laboratory to study urine and other pathologic specimens. In 1876 he traveled to Heidelberg and studied pathology and physiology, as well as clinical medicine. Upon his return, he was appointed to assist Delafield in the histology laboratory at Roosevelt Hospital in New York, where he studied many aspects of disease pathology and taught medical students as well. As one of his relatives observed, his chief attribute was "a talent for pure research combined with the ability to apply the latest results of scientific investigation to the amelioration of conditions which affect the welfare and happiness not merely of individuals, but of the human race as a whole" (Prudden, 1927).

Prudden clearly viewed himself as a medical scientist. He had obtained the best available training in science and medicine, both in the United States and in Europe. In an 1882 address to the medical students of the College of Physicians and Surgeons, he reflected the attitudes of many physicians toward science teaching within the medical school curriculum.

> We do not want or need to become zoologists or botanists or accomplished biologists in order to be good physicians . . . Special workers in these fields become far too engrossed in the minutiae of the theories which engage them and perhaps too deeply penetrated with the precision and exactness of the methods which they employ to permit of a devotion to that class of works and studies which the practical exercise of our profession demands . . . An elementary training in general biology and its methods of research greatly facilitates a thorough comprehension of the principles upon which much that is taught in medical school directly depends.
>
> The fact is that medical schools are simply technical schools, and it should be clearly understood that their first duty is to train physicians in the science and art of medicine as it exists today—to fit men to avert disease if possible, or to divert it to a favorable termination if they can, or, when the need comes to lead the hopeless patient as painlessly and gently as may be to his rest. (Prudden, 1882)

Prudden's attitude toward science was quite different from that of the practitioner of 1830 or 1840, for he had no fear that science would undermine the physician-patient relationship. In fact, Prudden encouraged his students to fully appreciate the work of those in the basic sciences, because there might be something useful from them in the future. But for the medical student who "may . . . fail of a thorough elementary training" in them, the sciences were viewed as unnecessary. Some familiarity with the basics was helpful for the practice of medicine, but it certainly was not essential.

This opinion was elaborated by D. W. Cathell, a Baltimore physician, in his popular 1882 book, *The Physician Himself, and What He Should Add to his Scientific Acquirements*. In this book of advice for young doctors, Cathell argued that to achieve success in the real world of medicine, a physician needed more than a thorough knowledge of the scientific aspects of the profession. He needed professional tact and business sagacity: "A well-polished manner and being moderately well versed in medicine will do much more good with the public than will special skills in histology, embryology or other 'ultrascientific' acquirements."

When new ideas in medicine were published, it was important that a practitioner's years of experience not be needlessly overthrown in their favor, Cathell warned his readers. "Do not be biased too strongly or quickly in favor of new or unsettled theories based on physiological, microscopical or chemical experiments. If you abandon the practical branches of medicine for histology, post-mortem researches, refined diagnostics and abstract reasoning, your usefulness as a physician will almost certainly diminish" (Cathell, 1882).

The physicians of the day did not know much about pathology or the mechanisms of drug action. These appeared to be irrelevant for daily practice. They worked hard, attended to their patients, and did fairly well by them. They knew enough to "get by" (Rothstein, 1987).

Medical School Education

The professional training of physicians after the Civil War did little to alter the concept of the physician as more small business entrepreneur than medical scientist. Medical schools uniformly offered practical, clinically oriented courses of study and generally required two years of lecture courses to qualify for graduation. Students sought a brief vocational education before beginning clinical practice (Rothstein, 1972). The pragmatic aspects of medicine were stressed, rather than theoretic ideas (Bonner, 1963). A survey in 1847 found that human dissection was studied in only 25 percent of the medical schools, and clinical demonstrations were not even offered in 30 percent of the nineteen schools. The yearly lecture series ranged from three to eight months in length (Kaufman, 1960). Basic science in the medical schools was presented in lecture format with "caution and kindness" to the students, who often had no previous experience in this area of study (Rothstein, 1987). They often found it impossible to understand how changes in normal structure and function resulted in disease (Rothstein, 1972).

Two factors inhibited the development of scientific education in the premedical and medical curricula. First, traditional opinion had been that the human organism was so different from other species that basic laws of biology could not apply. Hence, the study of science in general seemed of little import for the physician. But this opinion was slowly changing. One physician expressed the thought that "however mortifying it may be to the boosted superiority of the human race over all others of the animal creation, the physical man is an animal . . . and subject to the same immutable laws that govern the inferior species about him" (Griswold, 1881). Second, the perennial question continued to be argued as to whether experimental science would add anything to the ultimate aim of the practicing physician: to treat human disease (Warner, 1986).

The U.S. physicians who studied in Germany at this time became convinced that medicine was indeed a part of the science of biology. It sought to explain human disease in physical and chemical terms (Ludmerer, 1985). Virchow argued that "practical medicine shall become applied theoretical medicine, and theoretical medicine shall become pathological physiology . . . Pathological physiology takes its questions from pathological anatomy; partly from clinical medicine; it creates its answers partly from observations at the bedside; and thus is a part of clinical medicine, and part from experiments on animals. The experiment is the ultimate and highest resort in pathological physiology" (Harvey, 1981). Thus the laboratory sciences were coupled with the clinical sciences to understand mechanisms of human disease.

Efforts to improve the education and clinical training of U.S. medical students continued to come from physicians who had studied in Europe, a liberating experience that encouraged independent thought. These younger physicians were generally more receptive to new ideas than were their senior colleagues in medical schools and clinical practices, but they encountered significant resistance when they advanced the new scientific medicine as the way to practice in the United States. It took many years for the nation's intellectual climate to become receptive to their way of thinking (Bonner, 1963). Attempts to revise the medical school curriculum reflect the tension between the traditional and the new, scientific medical practitioners in the United States.

Harvard Medical School: The Physician as Scientist

The revision of the Harvard Medical School curriculum under President Charles Eliot exemplifies the way medical schools might be transformed from vocational to scientific institutions. Eliot had been trained as a chemist in Germany and was convinced that scientific training was necessary for future physicians. He believed that the application of scientific methods to human diseases would ultimately result in better diagnosis and treatment for suffering humanity. In 1869 the medical school curriculum consisted of two years of lectures without grading or written examinations. Each student was merely required to pass several brief oral examinations to graduate. It has been estimated that about 50 percent of the class of 1869 could not read (Bowers, 1976; Ludmerer, 1981).

The new curriculum advocated by Eliot after 1870 merged didactic and practical training into a three-year course. Elementary courses in the first year were followed by more complex lecture material and clinical training in the last two years. Written examinations were used in each course, and all courses had to be mastered for graduation (Huddle, 1991). Medical students became active participants in the educational process, working in chemistry, physiology, and pathology laboratories. This scheme was designed to teach them how to evaluate data not just to accept authoritative statements presented in lectures. The ability to think critically was felt to be more important than the traditional memorization of lists of facts, for Eliot believed that rapid advances in medicine would make such facts irrelevant within a few years of a student's graduation from medical school (Ludmerer, 1981).

Faculty opinion on the new curriculum was divided. J. C. White supported the new ideas and advocated a thorough scientific education for every medical student: "I would love to dispossess your minds of the too common belief that everything can be learned at the bedside; it is a fatal barrier to individual and national progress in medicine." He believed that "the greatest discoveries are waiting outside the sick room" in the research laboratory (Warner, 1986).

Senior faculty members H. J. Bigelow and Oliver Wendell Holmes vigorously opposed Eliot's new curriculum, fearing that more stringent requirements for medical school admission and graduation would prevent a young student with a natural genius for healing from obtaining a medical education (Starr, 1982). Bigelow believed that "most eminent men are in a large degree self-made . . . The material out of which philosophers are made is largely supplied from their intrinsic and determined will. Genius is latent with a strong driving power, whether versatile in all directions, or more profitably guided by tact or circumstance in one direction. You cannot create this talent or compel this taste. You may, indeed, give it opportunity, but you cannot force it" (Haller, 1981).

Bigelow was also concerned that the emphasis on science in the curriculum would divert attention from the primary role of the medical school. "Everything in medical instruction is to be made wholly subservient to the prevention and proper treatment of disease." He believed that the new science had little to offer therapeutics and argued that "the established rules of art are safer teaching than the speculations of science" (Warner, 1986). The new emphasis of the medical school on science, he feared, would alter the character of the physician from that of healer to scientific investigator (Huddle, 1991). "Experimental physiology leads away from safer and broader therapeutic views, and toward a local and exclusive action of chemistry and cells, uncertain grounds for students, from whom the results of larger and well-established medical experience is here the safest teaching." Bigelow proposed that the exercise of clinical judgment, not the application of scientific knowledge, was the essence of the therapist's task. The ideal physician, he thought, was not an expert in medical science but rather was possessed of "practical experience and sound judgment" (Warner, 1986). Bigelow was not opposed to all science, however, and in fact recognized the usefulness of clear thinking as applied to clinical observations. He noted that one positive outcome of the curricular debate was that "the barren fields of speculative hypothesis and arbitrary assertion have thus been fairly replaced by the precise methods of induction from observed phenomena" (Huddle, 1991).

The emphasis on critical thinking became an ever-increasing component of medical education toward the end of the nineteenth century. Memorizing purported facts was less important than attempting to understand basic biologic phenomena and testing proposed ideas with laboratory experiment and clinical experience. An appreciation of basic science was no longer viewed as strictly esoteric. As physicians gradually came to view themselves as scientists who applied this basic knowledge to the practice of medicine, their view of the role of heredity as an important factor in causing human disease changed.

THE MEDICAL LITERATURE: EMBRYOLOGY AND THE HEREDITY-EVOLUTION DEBATE

The growing level of interest in biology in general and heredity in particular becomes evident upon analysis of the medical literature of this era. The monthly *Index Medicus* had compiled references from the major medical journals since 1879. In volume 6 for 1885, the word *heredity* was first used for cataloging. Several articles each year thereafter were listed under the category "Heredity and Disease." When the massive *Index Catalogue of the Library of the Surgeon General's Office* was published in 1885, the topic "Heredity and Disease" comprised five pages with dozens of references in its sixth volume (Billings, 1885).

Although the results of research in experimental embryology and heredity were regularly reported in the U.S. medical literature after the Civil War, their relevance to medical practice first demanded general agreement among physicians that biologic laws for hereditary transmission were the same in humans as in other species (Hereditary transmission, 1875). A particulate model for heredity was widely discussed in the medical literature after the theories of Spencer and Darwin had been published. The model held that particles of inheritance, or "plastidules," were bundles of matter transmitted via egg and sperm to form the fertilized ovum. These physiologic units then guided the development of the embryo (ibid., 1875; Elsberg, 1882; Dolan, 1888–89). The actual mechanism by which these units were expressed was not understood. "There may be circumstances of which we are now ignorant which may cause the exhaustion or diminution, or lessen the influence of any particular plastidule" (Elsberg, 1879). In fact, this was a chemical model for inheritance: these transmitted factors were viewed as chemical entities comprised of atoms. The union of the factors from the two parents might produce a new combination of atoms which then would result in a unique trait in the progeny. This process might be beneficial or detrimental. In a family with hemophilia, one physician suggested, "We may suppose that two or more harmless factors were so united in marriage as to produce in their descendants this strange disease" (Rachford, 1892). Another commented that, given the incredible complexity of the human being, it was truly wonderful that more errors in heredity and development did not occur (Hoke, 1889).

Professional biologists were invited to medical society meetings after 1880 to inform physicians of the latest developments in heredity as they applied to clinical work. The embryologist from Harvard Medical School, for instance, C. S. Minot, prepared a detailed discussion of the formation of the human sperm based on his own laboratory research. He presented data that showed that the chromatin within the cell nucleus was "the material substratum of hereditary transmission . . . Chromatin is the essential factor in the fruition of heredity" (Minot, 1886a). The expression of this hereditary potential during embryonic development resulted in the birth of each unique human being. "We are in fact forced to the hypothesis that the chromatin is the physical basis in which the formative forces of the organism is actually situated, and that the reason why children are like their parents is that they share of some of the same kind of, actually some of the very same chromatin, as their parents and therefore the same forms are reproduced because the form-regulating substance is identical. Variations occur because there is chromatin from two parents, and the two sets of chromatin may differ in potency" (Minot, 1886b).

When H. F. Osborn of Princeton University discussed human heredity with a medical audience several years later, he agreed that "the vehicle of heredity" must reside in the chromatin of the cell nucleus. He described the organization of the chromatin into rodlike structures during mitosis or cell division and labeled these "chromasomes." Each cell of an organism appeared to have the same number of "chromasomes." During mitosis, each rod split longitudinally, and one portion was carried into each of the two resulting daughter cells. The production of egg and sperm, however, required a different type of cell division. This involved meiosis, a reduction division that resulted in egg or sperm with one-half the original complement of "chromasomes." Fertilization of the ovum involved the actual fusion of the nuclei from the egg and the sperm. The uniqueness of each human being was the result of this reduction division followed by the union of the hereditary factors from both the mother and the father. He argued that this mechanism "offered an infinite number of combinations for selection to operate upon" (Osborn, 1892).

J. P. McMurrich, a biologist from the University of Cincinnati, emphasized the same points in his discussion of heredity at the Ohio Medical Society. The chromatin of the germ cell nucleus was clearly the "material bearer of heredity." The molecules of the chromatin corresponded to the various structures of the parent organisms (McMurrich, 1894).

After 1890 many physicians accepted the concept of particulate inheritance as a logical explanation for the transmission of hereditary characters, a circumstance reflected in three medical articles in 1897 which reviewed the mechanics of inheritance as it was then understood. The chromatin in the cell nucleus, generally believed to be the hereditary substance, could be viewed as a thread consisting of "thousands of vital units." During mitosis the number of units was equally divided and then segregated into the two daughter cells. The reduction division, or meiosis, that produced the egg and sperm decreased the amount of "accumulated ancestral hereditary" material. Union of egg and sperm then restored the absolute amount to normal. The expression of the hereditary units depended upon the potency of the material derived from the mother and father, a process described in evolutionary terms as a "struggle of heritages" within the cell nucleus, which eventually resulted in the unique structure of the newborn infant (Gordan, 1897; James, 1897; Shute, 1897).

Heredity was always viewed by physicians of this era both as an event at fertilization and as a process that unfolded during embryonic development. It involved a tension between the immutable features of the human race and the subtle variations that comprised each individual (Pallen, 1856). This belief reflected the controversy in biology about the distinction between heredity and evolution. One speaker before a local medical society remarked that "students of biology have ranged themselves in hostile armies over the various theories of heredity" (Walker, 1897–98). Gasser believed that one could not really isolate the two poles of the discussion. Weismann's model, he thought, provided a good mechanism for heredity, but it was unclear how evolutionary selection could act on the germplasm, as Weismann defined it. The Lamarkian and Darwinian theories of evolution also made sense but did little to explain hereditary transmission. Gasser proposed that each union of the germplasm at fertilization involved a rearrangement of matter. Because the "organizing tendencies of the parents" in the germplasm were in conflict, the newborn baby was the result of a "struggle for the supremacy of expression." But there appeared to be more to embryonic development than the mere unfolding of innate tendencies. Life itself, Gasser believed, was dependent upon the interaction between inborn and environmental forces. This conflict resulted in the structure of each individual being (Gasser, 1895).

The U.S. physicians of the day were always practical. Their interest in theories of heredity was limited to the improvement that these might make in the care rendered to patients. A speaker before the Maine Medical Association discussed heredity from this practical viewpoint, "not to rehearse theoretical information, nor to introduce new scientific schemes and theories alone" (Fuller, 1887). Another speaker, this time at the Connecticut Medical Society, suggested that physicians should consider "the pathologic relations of heredity alone, leaving the physiological concerns to the scientists in which to revel" (Sheets, 1889). The task for physicians was to glean useful information from these theories of heredity so that they might better understand the nature of human disease.

CLINICAL PRACTICE AND RESEARCH ON HEREDITY, 1880 TO 1900: REEVALUATING "LIKE BEGETS LIKE"

The physicians of this era sought explanations for patterns of disease that fit the observations that they made in daily practice. Although aware of different theories of inheritance under investigation by biologists, they were not inclined to change the rules for inheritance that worked for them, the direct, indirect, and collateral modes of inheritance outlined by Ribot.

Although the notion of inheritance as "like begets like" was still affirmed by physicians in the last quarter of the nineteenth century (Wright, 1880–81), after 1880 they expressed less certainty that disease itself was hereditary and became increasingly convinced that only predisposition to disease might be inherited (Iles, 1878; Clark, 1880–81; Couch, 1880–81; Coulter, 1896; Richardson, 1896). In general, offspring resembled their parents in many respects: physical, mental, and moral traits continued to be accepted as hereditary (Chancellor, 1887), with their expression dependent upon specific triggers from the environment (Von Schaick, 1888; Christison, 1895; Gordan, 1897; Crandall, 1897).

Specific diseases that had been labeled as hereditary—and hence inevitable—in the past were reinterpreted in the light of this model. Epilepsy certainly appeared in successive generations of many families, but not all offspring were affected. It was regarded now as a reflection of general neurologic degeneration. In various members of a family the predisposition might be manifested as neuropathy (dysfunction of the peripheral nervous system), epilepsy, or severe mental illness, depending upon environmental influences (Mann, 1883; Hamilton, 1886; Temkin, 1971). Mann called it a "family neurosis" that manifested itself in different generations as mental retardation, insanity, inebriety, or epilepsy (Mann, 1883). Likewise, some trigger activated the latent predisposition toward the bleeding diathesis hemophilia, which affected some but not all males in families in which the unaffected females appeared to transmit the trait (Hopkins, 1880–81). Pseudohypertrophic muscular paralysis followed a similar pattern. Bridge thought that it might be hereditary to a "slight extent," but it should more properly be called a diathetic disorder peculiar to males in certain families. The actual expression of dysfunction once again depended upon some external inciting event (Bridge, 1885).

Evidence also accumulated that showed the influence of heredity in the development of different types of cancer. Between 15 and 20 percent of liver cancer patients were found to have an affected ancestor. One investigator commented that "heredity is the most important factor" (Barthelow, 1885). William Welch, professor of pathology at Johns Hopkins, reviewed 1,744 cases of stomach cancer and found affected relatives in 14 percent. He noted that the disease in such cases developed at an earlier age than in those without a family history of the same disorder (Welch, 1885). Another faculty member from Johns Hopkins, William Osler, reported a family in which three successive generations were afflicted with leukemia (Osler, 1892).

A summary of the effect of heredity in familial cases of uterine cancer described the opinion of many physicians from this era:

> If there is anything in the idea of heredity as a causative influence, it must be rather through physiological similitude of children to parents than the transference of taint from the former to the latter. If cancer is a degeneration of tissue, as the effect of a law that organs in certain individuals undergo dissolution at a certain age, we can understand that the child may inherit such physiological effects from the mother. The cell formation of the organs of the child will be capable of reaching the same period at which the disease was developed in the mother, when the normal histological changes will be interrupted and dissolution begin. In this view of the subject, the child would by action of its organization inherit the mode of dying evinced in the mother. (Byford, 1886)

The predisposition to cancer rather than the disease itself was passed from one generation to the next.

The problem of familial tuberculosis exemplified the difficulties encountered by physicians as they tried to discover the causes of specific diseases. Osler reported a family in which six successive generations were afflicted by pulmonary tuberculosis. Although he believed that the disease could be inherited, he could not distinguish the actions of heredity from those of infection as causes of the disease (Osler, 1892).

A more restricted definition of what constituted heredity developed among U.S. physicians at this time. Heredity came to mean *direct* heredity as defined by Ribot, the "conveyance of a definite morbid taint

Table 2.1. Direct Heredity Reported in Huntington
Chorea, 1885–1900

Generations Affected	Pedigree Provided in Study	Reference
4	No	King, 1885
3	No	King, 1885–86
3	No	Diller, 1889
2	No	Hay, 1889–90
4	No	King, 1889
5	No	Sinkler, 1889
4	No	Hay, 1891
4 or 5	No	Stephens, 1892
2	No	Osler, 1893
4	No	Butler, 1894
2	No	Osler, 1894a
2	No	Brush, 1895
5	Yes	Dana, 1895
3	Yes	Collins, 1898a
3	No	Collins, 1898b
4	Yes	Hallock, 1898
2	No	Smith, 1898
4	Yes	Berry, 1900
3	Yes	Sinkler and Pearce, 1900

from one generation to another" (Witter, 1887). The schematic representation of hereditary transmission by the pedigree or family tree became more commonly used by practitioners, but almost exclusively to represent heredity of this direct type. The quintessential example of direct heredity was generally believed to be Huntington chorea, in which it was agreed that "a more strongly marked heredity than is seen in almost any other disease is prominent as a symptom" (Phelps, 1892). In nineteen reports of families with chorea from this era, five provided pedigrees in the original paper (table 2.1). That the disorder could be represented in this manner implied that it was truly an example of heredity at work. Few comments about heredity were deemed necessary when such a pedigree could be drawn to demonstrate direct transmission of the trait.

Analysis of the familial nature of another neurologic disease produced quite different findings. Friedreich ataxia often affected several children in one generation of a family. Progressive incoordination and weakness of the muscles resulted in poor coordination and speech difficulties throughout adulthood (Spitzka, 1885; Osler, 1892). Data

Table 2.2.   Heredity Reported in Friedreich Ataxia, 1885–1900

Type of Occurrence	Pedigree Provided in Study	Reference
Single case	No	Sinkler, 1885
5 siblings	No	Smith, 1885
Direct	No	Fellows, 1886
2 siblings; 2 siblings; 4 siblings	No	Griffith, 1888
3 siblings	No	Shattuck, 1888
5 siblings	No	Wells, 1888
2 siblings	No	Fry, 1893
2 siblings	No	Burr, 1893–94
Collateral: 5 siblings, 1 cousin	Yes	MacKay, 1894
Single case	No	MacKenzie, 1894
Single case	Yes	Nammock, 1894
2 siblings	No	Zenner, 1894
4 siblings	No	Halbert, 1895
4 siblings	No	Small, 1895
3 siblings	No	Coe, 1896
2 siblings	No	Fry, 1896
2 siblings	No	Riggs, 1896
Direct and collateral: 3 siblings, aunt, uncle	No	Brower, 1897
Collateral: 2 siblings, 2 cousins	No	Kellogg, 1898
2 siblings	No	Starr, 1898

from twenty reports of this disorder from 1880 to 1900 showed that ataxia was clearly a family disease (table 2.2). It had been called hereditary, but cases of direct transmission were rare (Smith, 1885). Parents were commonly unaffected, although the cases reported by Brower involved an affected father and four of his children. Much more commonly, siblings of the same generation had the disease, while parents and other relatives were quite healthy. More distant relatives in collateral branches of the family also could have the same symptoms.

Only two of the reports of ataxia presented family pedigrees. The first (MacKay 1894) was written by a British physician. In the second, Nammock used a pedigree but argued that the disorder was not hereditary: "It would seem desirable to report other non-hereditary cases, in order to eradicate the fixed idea that Friedreich's ataxia is necessarily a family disease" (Nammock, 1894). The absence of direct parent-to-child transmission argued strongly against the hereditary nature of this trait:

> These cases tend to show that the term "hereditary" ataxia is a misnomer and misleading. (Shattuck, 1888)

> As to heredity, this affection is more properly a family disease, not necessarily hereditary, i.e., several brothers and sisters may be affected by it, while there is no trace of it among the ancestry. (Riggs, 1896)

> Previous reports of "hereditary" ataxia . . . are not strictly speaking hereditary, for neither the parents nor any of the ancestors, as far as known, were ataxic. (Sinkler, 1885)

> It is probable that all sufferers of the disease inherit from some ancestors, near or far, a tendency to degenerative processes, yet to call it hereditary we must regard the term in its broadest sense. (Small, 1895)

> The disease then is rarely hereditary in the strict acceptation of the term. (Smith, 1888)

Similar observations were made in familial cases of albinism, a disorder in which skin pigment is absent (A Curious Case, 1884; Jennings, 1890); cataracts (Powers, 1892); and xeroderma pigmentosum, a skin disorder with chronic ulceration and malignancy (Broyton, 1892–93). Many siblings in one generation of such families often were affected, but parents and other relatives were not. It did not make sense to label these hereditary diseases because there was no evidence of direct transmission of the predisposition.

In the last decade of the nineteenth century the working definition of heredity as "like begets like" was reevaluated. Heredity came to be understood as the direct transmission of a character from parent to child. Twenty-eight additional references from the literature used pedigrees to outline the inheritance of various human traits, and in each instance the ability to demonstrate direct parent-to-child transmission was accepted as strong evidence for heredity as a factor (table 2.3). The only exceptions to this rule involved the special cases in which healthy females appeared to transmit disease traits to their sons. These examples were also accepted as evidence of heredity at work in producing symptoms of specific diseases.

In 1892 the ophthalmologist C. A. Wood made an observation that would be used in the next decade to demonstrate that heredity did in fact function in humans as it did in other animal species. He reported a

Table 2.3.   Heredity Studies Providing Pedigrees, 1880–1900

Trait	Reference
Direct heredity	
Optic atrophy	Norris, 1880–84, 1882
Coloboma irides	DeBeck, 1886
Polydactyly	LeConte, 1886
Tremor	Dana, 1887
Glaucoma	Howe, 1887
	Harlan, 1898
Muscular atrophy	Harrington, 1887–88
Cataract	Wilson, 1891
Ataxia	Brown, 1891–92
	Neff, 1894–95
Tuberculosis	Osler, 1892
Nystagmus	Wood, 1892
Lymphedema	Milroy, 1892–93
Iridemia	DeBeck, 1894
Deafness	Neff, 1895–96
Tylosis palmae et plantae	Ballantyne and Elder, 1896
Ectopia lentis	Miles, 1896
Spastic paraplegia	Bayley, 1897
Parakeratosis	Gilchrist, 1897
Myotonia congenita	Clemesha, 1897–98
Periodic paralysis	Taylor, 1898
Microcornea	DeBeck, 1900
Hemophilia	Steiner, 1900
Sex-linked heredity	
Hemophilia	Osler, 1885
Optic atrophy	Gould, 1893c
	Posey, 1898
Cryptorchidism	Homan, 1898

family in which two traits appeared to be inherited together. Individuals in three generations had hereditary nystagmus, transmitted directly from parent to child. Both sexes were affected, but those with nystagmus were brunette, while those with normal eye movements were blonde (Wood, 1892). Examples of such coupled characters would be found in many other species after 1900 and were explained on the model that hereditary units for several physical traits were physically bound to the same chromosome in the cells of the body. As such, they would be transmitted together from one generation to the next. This argument for genetic linkage was eventually accepted by many in the field of hereditary research to prove the existence of physical units for heredity. After the discovery of Mendel's work in 1900, the cosegregation of these units could be explained as a consequence of the gene theory of inheritance.

Physicians now clearly understood heredity to be a process rather than an event. A speaker at the American Pediatric Association in 1897 defined heredity as "the tendency of an organism to develop in the likeness of its progenitors" (Crandall, 1897). There appeared to be a powerful force on the part of nature to preserve uniformity within the species. Deviations from the common type were often detrimental to health. Uniformity, not variability, was the major role for heredity in development.

The notion of hereditary disease also fell into disfavor among physicians because it appeared to place the healing profession in a helpless position. If disease was hereditary and therefore inevitable, fatalism and despair would prevent the physician from even attempting to treat the affected patient (Von Schaick, 1888). This threat to professional integrity did nothing to encourage physicians to accord the concept of hereditary disease a great deal of attention. In fact, it encouraged them to avoid the issue as much as possible and to look for causes of disease which might be more amenable to their ministrations and hence less threatening to their professional existence.

The Dilemma of Consanguinity and Heredity

Cases of the blending of blood lines with common ancestors, or consanguinity, were believed to provide data on the heredity of diatheses for human disease. In certain families, consanguinity resulted in no discernible problems among the offspring. But commonly defects did occur: "Whatever elements of constitutional weakness or deficiency may exist undeveloped in the parents, being probably similar in the two, owing to the community of blood, would be much more likely to appear in the child, than if the parents were unrelated" (Ray, 1862). Both biologists and physicians undertook extensive analyses of inbred families to determine the nature and extent of both positive and deleterious features.

In 1858 Bemiss reported his study of 873 consanguineous marriages. He estimated that of all impaired children, the children of cousin parents comprised 13 percent of offspring who were deaf, 8 percent of those who were blind, 13 percent of those who were mentally retarded, and 2 percent of those with mental disorders in his sample (Bemiss, 1858). Other large surveys found even more impaired offspring when the parents were closely related. Two studies on the fre-

Table 2.4. Impairedness of Offspring of Consanguineous
Marriages, 1888–1891

	Percent Impaired Offspring	
Relatedness of Parents	Dolan	Lydston
Third cousins	41	
Second cousins	42	
First cousins	67	48
Uncle × niece	81	80
Brother × sister	97	94

Source: Data from Dolan, 1888–89; Lydston, 1890–91.

quency of impaired children born of related parents found more impaired offspring as parents were more closely related (table 2.4).

This type of information was interpreted to mean that the admix-
ture of two lines from the same family tended to exaggerate predispositions in the offspring (Outter, 1870; McKee, 1888). The closer the
parents were to a common ancestor, the more likely they were to possess a common impairment, which could then be transmitted to their
children. Numerous studies reported before 1900 examined the occurrence of specific diseases within inbred families in the United States
(table 2.5).

Not all marriages of relatives, however, produced impaired offspring. Other studies found little evidence of malformation, at least in
marriages between cousins (Withington, 1885; McKee, 1886). Alexander Graham Bell examined marriage statistics among people who
were deaf and found that 33 percent had affected relatives. A nonhearing parent who chose a hearing mate had a 20 percent likelihood of
producing a child who was deaf, but two deaf parents on average produced an affected child in only 10 percent of cases. He noted that
hearing people who married nonhearing partners often came from
families with other nonhearing relatives. He interpreted these facts as
suggesting that the parents might carry some predisposition for deafness which could be transmitted and exaggerated by marriage to an
affected partner (Bell, 1884). Bell's colleague E. A. Fay recognized that
deafness was not a single disease but rather a symptom that could result
from many different causes. If one deaf parent had a particular disorder and the other deaf parent had an unrelated anomaly, there should
be no particular intensification of the predisposition to deafness in
their offspring. Theirs was not really a marriage of "like with like" (Fay,
1898).

Table 2.5. Disorders Common in Consanguineous
Families, 1860–1900

Disorder	Reference
Retinitis pigmentosa	Webster, 1878
	Ayres, 1886
	Coleman, 1889
	Lydston, 1890–91
	Belt, 1896
	Debeck, 1897
Blindness	Layton, 1882
Clubfoot	Layton, 1882
Epilepsy	Layton, 1882
Insanity	Layton, 1882
Polydactyly	Layton, 1882
Deafness	Bell, 1884
	Fay, 1898
Cataract	Alt, 1887
Spina bifida	Holt, 1887
Aniridia	Jennings, 1890
	Lydston, 1890–91
Albinism	Jennings, 1890
	Lydston, 1890–91
Aphakia	Lydston, 1890–91
Coloboma iridis	Lydston, 1890–91
Microphthalmia	Lydston, 1890–91
Xeroderma pigmentosum	Hutchins, 1893
Amaurotic family idiocy	Carter, 1894
Friedreich ataxia	MacKay, 1894

But if heredity was defined as the direct transmission of a characteristic from parent to child, its role in consanguinity was confusing.
With rare exceptions, the parents of children with xeroderma or albinism were unimpaired. The parents may have transmitted a character
that was latent in themselves, but how could the union of such factors
from both parents result in the appearance of children with traits
unknown in previous generations of the family?

In the last decade of the century the working definition of heredity
as "like begets like" was thus reevaluated. Physicians encountered a
number of disorders that affected several brothers or sisters in a family
in which no previous relatives had been afflicted. Cousins and other
relatives in collateral branches were occasionally affected (Brower,
1897; Kellog, 1898), but did such cases really constitute hereditary
diseases?

The Debate over Inheritance of Acquired Features

The sudden appearance of a disorder in the children of families with no affected relatives required an explanation. The process of human development was believed to generally produce a uniform type of individual (Hoke, 1889), but more precisely it involved a balance between many immutable forces and some degree of variation between individuals of the species (Pallen, 1856). Physicians at midcentury shared with biologists of the era the general opinion that acquired features could become hereditary (Outter, 1870). Reproduction was viewed as the blending of features from both parents. Patterns of use, disease, or injury might alter the hereditary factors in one parent and result in the transmission of acquired features via inheritance.* One of the major proponents of this theory was the German biologist Ernst Haeckel. He believed that both morphologic and functional features acquired by an individual as a result of the external environment could be transmitted to offspring, citing examples of the inheritance of albinism, polydactyly, birthmarks, and injuries. That acquired features could be inherited was an important truth, in his opinion, because it generated significant variation in the species and was intimately connected with the process of evolution (Haeckel, 1866).

Spencer was also convinced that the inheritance of acquired features was necessary for the theory of evolution to make sense. He cited examples in humans of how such variation could occur, among them a white woman who had children first by a black man and subsequently by a white man. The children with the white father were reported to have traces of black features. Spencer argued that this observation indicated that the traits of one father had altered the mother's germ cells, as shown by the subsequent progeny. If the germ cells could be altered, there was no apparent obstacle to the transmission of acquired characters from one generation to the next (Spencer, 1893).

Darwin generally accepted the inheritance of acquired features as a source of variation which was an integral part of his theory of evolution by natural selection. Toward the end of his career, however, he became progressively more troubled by the entire issue. He noted that among a population living under identical conditions, variation appeared in only a few individuals and not the entire group. He came to conclude that the nature of variation depended to a small extent on

---

*Several large-scale reviews have examined the history of this topic in nineteenth-century biology (Kammerer, 1924; Zirkle, 1946; Blacher, 1982; Bowler, 1989).

external conditions and more on the constitution of the individual organism (Darwin, 1898).

Leading American biologists agreed on the importance of the heredity of acquired features for the general theory of evolutionary change. E. D. Cope (Baller, 1976) and H. F. Osborn (Osborn, 1889) were public advocates of this position. Osborn in particular was concerned about the absence of a satisfactory mechanism to explain how such variation was transmitted from one generation to another.

Other biologists raised even more serious questions about the validity of the notion. Wilhelm His in Germany noted that certain human races had practiced circumcision for thousands of years and yet males in those communities were still born intact (His, 1874). By far the leading opponent of the heredity of acquired features was A. Weismann. His review of the scientific data on the topic labeled the whole notion "obscure" (Weismann, 1891–92). His concept of the germplasm as separate from the somatoplasm implied that hereditary factors within the germplasm were generally unaffected by external factors in the body of the organism, although he did not deny the possibility of change within the germplasm. External factors, he believed, could produce "minute alterations in the molecular structure of the germplasm." As this material was transmitted from one generation to the next, such changes would become hereditary (Weismann, 1891–92).

Because biology in the late nineteenth century had become primarily an experimental science, speculation or natural history observations were no longer acceptable foundations for new theories. Testable hypotheses that could be verified by experimentation were regarded as the source of facts in science. Galton attempted to test the theory of the heredity of acquired features by transfusing blood from a strain of rabbits of one color into that of another color, then breeding the recipients. After several generations there was no evidence of coat color change in the offspring (Galton, 1871, 1876). Weismann attempted to produce a race of tailless mice by amputating the appendage and inbreeding the animals. He eventually collected data on twenty-two generations, but none of the 1,592 offspring was born with a shortened tail (Weismann, 1913). On the other hand, Lockwood reported that he had clipped the tails of rats and inbred them for seven generations to establish a tailless strain (Talbot, 1898). Shidell also claimed to have produced a breed of tailless mice after clipping and inbreeding ninety-six generations of animals; given the number of years such breeding would have required, Woodruff argued, such results were impossible (Woodruff, 1900).

The most influential proponent of experimentally induced hereditary variation was C. E. Brown-Sequard. Beginning in the 1870s, he and his colleague Dupuy in France reported numerous studies in guinea pigs in which the sciatic nerve or specific columns of the spinal cord were transected. The offspring sometimes demonstrated neurologic dysfunction, such as epilepticlike seizures or deformities of the ears, eyes, or limbs. This was widely regarded as evidence for the inheritance of acquired neurological disease (Brown-Sequard, 1875; Dupuy, 1877).

Toward the end of the century, however, attempts to reproduce these studies yielded highly variable results and raised serious questions about their validity. For example, Hill cut the cervical sympathetic nerves in guinea pigs to produce drooping eyelids. Inbreeding of these animals failed to yield any progeny with similar traits (Irwell, 1902a). Numerous other reports of attempts to experimentally induce variation that could then become hereditary also resulted in only negative results (Blacher, 1982).

In contrast were observations by U.S. physicians which provided many examples of acquired features that subsequently appeared in the sons and daughters of the affected individual. The apparent direct transmission fit their understanding of how heredity worked. One physician commented that "no theme opens a wider field for speculation than heredity" (Gibney, 1876). Two vignettes illustrate this line of reasoning in U.S. medicine from 1850 onward. In the first, a woman was kicked in the nose by a cow. Her children then had frequent nosebleeds (Pallen, 1856). The second concerns a man who fought in the Civil War and pressed his musket to his shoulder over a prolonged period of time. An exostosis developed there, and the trait was passed to his children and grandchildren (Gibney, 1876).

In spite of the skepticism noted previously of U.S. physicians toward medicine as an experimental science, they were aware of the controversy in biologic circles over the experimental evidence for and against the heredity of acquired features. Their general belief in the reality of this phenomenon was bolstered by the well-publicized experiments of Brown-Sequard and Dupuy (Iutzi, 1882). Weismann's arguments against the heredity of acquired characters were not well received in U.S. medicine, where opinion was strong that the germplasm could be altered by environmental factors rather regularly to provide a mechanism for the transmission of acquired features from parent to child (Richardson, 1890–91; Nunn, 1895; Heredity of acquired characteristics, 1897). At the end of the century the great majority of U.S. physicians accepted as fact that "the acquired ailment of the parent becomes the infirmity of the offspring" (Carlow, 1901). The presence or absence of experimental evidence counted much less than the accumulated experience of clinicians in daily practice (Gordan, 1897). A professor of anatomy at the University of Oregon Medical School commented that "no one . . . is more capable of criticizing any questions of heredity than the medical profession, whose education makes them capable of observation, and whose profession gives them every opportunity for such observation" (Nunn, 1895).

Impressions during Pregnancy

There was general agreement among physicians before 1890 that the presence of a defect or disorder at birth indicated that it was hereditary in nature (Cleveland, 1877). Heredity was viewed as a process occurring throughout embryonic development, not only as an event associated with fertilization. Therefore, it was plausible that external factors could alter the course of normal development and produce variations that were evident at birth (Hartshorne, 1885). These "maternal impressions" were perceived as a special type of hereditary acquired character.

Numerous examples of defects noted after birth were correlated with events that occurred during human gestation. The birth of an anencephalic child who looked like a monkey was attributed to the mother's being frightened by an organ grinder's monkey during the third month of the pregnancy (Hope, 1879–80); another mother, frightened by an alligator at four months of pregnancy, produced a child with congenital ichthyosis—thickened, scaly skin (Fox, 1884); a baby with an eye disorder was born to a mother who was very anxious during the entire pregnancy and spent a great deal of time crying and rubbing her eyes (Amick, 1884); when another woman was two months pregnant, a cat scratched the skin below her left nipple and left a "deep impression on her mind." Her daughter was born with an extra left nipple (Coe, 1888); a child was born who looked exactly like a man who had been seated opposite the mother at a dinner party early in the pregnancy (and was not her husband) (Danforth, 1888–89); and at two months of pregnancy a woman was frightened by some dogs who were tearing the ears off a pig. Her child was born with a ragged external ear (Briggs, 1885–86).

Based upon this type of evidence, the relentless law of maternal impressions was widely accepted as fact by physicians. Not until the end of the century was it recognized that most cases of alleged impression occurred at times during the pregnancy when the major organ systems had already been formed. Defects in embryogenesis certainly did occur, but often these were early in the pregnancy, before the mother was even aware of her gravid state (Christian, 1889; Sherwood, 1899). By the last decade of the century, understanding of human embryology had developed to the point where a reassessment regarding maternal impressions seemed in order. Briggs suggested in 1885 that the issue was subject to scientific scrutiny and should be investigated with "more research and less preaching" (Briggs, 1885–86).

Implications of Heredity: Physicians as Social Leaders

While controversy among physicians continued on the question of the heredity of disease, there was wide agreement on the important role played by heredity in emotional and moral characteristics. U.S. physicians were typically solid members of the middle class and shared the attributes and opinions of their counterparts, convinced that U.S. society was to be the model for future civilization around the world (Furnas, 1969). The key to progress was understood to be the application of science and technology to the problems of society. Spencer's concept of social Darwinism became a potent force in the nation's social thought. The "survival of the fittest" concept from biology seemed to provide a rationale for the national passions of competition, avarice, and personal success, which were viewed as forward-looking social traits (Ludmerer, 1972).

Scientific naturalism espoused the idea that personal character was determined by the physical constitution of one's brain. The hereditarian view of the nature-versus-nurture controversy (Cravens, 1978) implied that moral and emotional characters would be susceptible to scientific study and eventual modification. Hence biology, not sociology, held the key to social progress (Bowler, 1989). Since contemporary U.S. society was accepted as representing the ideal, its members had a God-given right and responsibility to realize their divine mandate.

Any perceived threat to the status quo upset the establishment, including physicians. Toward the end of the century the vast numbers of immigrants from the Orient and southern and eastern Europe, with their foreignness and large family size, posed such a threat. There was general concern that they were felons and illiterate people (Furnas, 1969); they were called "nervous, anxious, and curious," as opposed to their more placid neighbors who probably had stayed at home in the old country (Millikin, 1881). These views also implied a social hierarchy of human races. Some were believed to be inherently unfit, while the Caucasians were viewed as superior (Bowler, 1989). And these distinctions were felt to be transmitted by heredity from one generation to the next. "The peculiarities of mental constitution that make the Caucasian the most progressive race are handed down by inheritance just as truly as the color of the skin and the shape of the skull" (Stoller, 1890).

Galton, who had studied the nature of hereditary factors in human populations as early as 1870, coined the word *eugenics* to describe a science of selective breeding which could be applied to the improvement of mankind (Galton, 1883). If civilization was to progress it was necessary to control the number of unfit offspring, for these would eventually become burdens on society. There was genuine fear that the rapid increase in the number of social misfits was the result of indiscriminate breeding of the disadvantaged classes. The millions of dollars required for their care depleted the public resources available for the future expansion of society at large (Garver, 1895; Reynolds, 1896).

U.S. physicians, convinced that the primary cause of such inferior people was hereditary predisposition, noted that identifiable social traits recurred in successive generations of particular families (table 2.6). As early as 1875 one physician commented that "it is certain from observation that man's mental status and moral stature are hereditary" (Hereditary transmission, 1875). Another concluded, "Much of vice, pauperism, idiocy and insanity is hereditary" (Wright, 1880–81). There also appeared to be a consistent hereditary predisposition to theft, murder, and suicide (Layton, 1882).

Table 2.6.   Hereditary Predisposition and
Social Character from Reports before 1900

Character Trait	Reference
Pauperism	Parker, 1877
Drunkenness	Millikin, 1881
Hysteria	"
Insanity	"
Vagrancy	"
Imbecility	Layton, 1882
Murder	"
Suicide	"
Theft	"
Dementia	Garver, 1895
Mania	"
Melancholia	"

The strong deterministic opinions of U.S. physicians were reflected in their analysis of how heredity should be applied to this problem. One believed that it was now time to apply the laws of heredity to "solve the problems of crime and pauperism" (Parker, 1877). The logical consequence of this belief was that the laws of heredity comprised the "fountainhead of all social improvement" (Aller, 1885–86).

The general agreement on the importance of brain structure for proper human conduct suggested a minor role for external influences on human behavior (McKim, 1899). No one denied that exercise, good food, fresh air, and education could modify the outcome of hereditary disorganization (Peters, 1879; Witter, 1887; Spratling, 1894), but the overwhelming cause of society's ills was felt to be nature, not nurture.

Paul Starr has examined the gradual acceptance of the physician as the expert in matters of health by the U.S. public at this time. Rapid improvement in general health did occur after 1890 as the frequency of many infectious diseases declined through the application of medical science. The physician was perceived as having special technical competence in this area and was recognized as an increasingly valuable member of society (Starr, 1982). In the area of social ills the physician was urged to take action to eradicate hereditary defects to improve humankind in general. If only the morbid tendencies could be removed, then surely a moral and social person would result (Snediger, 1896).

Physicians were entrusted with the power to encourage the marriages of certain individuals and to dissuade other, less desirable couples from marrying. They were urged to educate people in the choice of proper marriage partners. In this way they would act as animal breeders in facilitating the union of the best stock (Snediger, 1896). Certainly there would be some unhappiness in a few lines, but it would be far better to advise against any marriage that would add to the ever-increasing number of undesirables appearing in U.S. society (Williams, 1880; Millikin, 1881; Holloway, 1885; Starr, 1982).

Various approaches were suggested to implement this social policy. Some physicians proposed isolated colonies for persons with particular hereditary traits such as epilepsy. These establishments could be "humane, curative and economical." With good food and fresh air, the general health of the inmates should improve and the colony could become self-sufficient. Strict segregation of the sexes was important to prevent cohabitation and any increase in undesirable offspring (Spratling, 1894).

Stricter marriage laws were also proposed as a part of the public education program to discourage unions between kindreds with socially unacceptable traits (Reynolds, 1896). For mentally ill people or those with wicked behavior, sterilization was proposed to reduce the chances of affected offspring (Hereditary transmission, 1875; Garver, 1895; Reynolds, 1896). In cases where nothing else would stem the tide of social misfits, euthanasia—a "gentle and painless death"—was advised as the ultimate solution to society's problems (McKim, 1899).

With U.S. physicians of this era expressing a consistent belief in the hereditary nature of many social ills, it is small wonder that the logical "cure" involved the prevention of reproduction in affected families. Surely the next generation would be more fit, and its members more able to enjoy the fruits of progressive society.

ON THE CAUSES OF DISEASE: NATURE VERSUS NURTURE

Although physicians generally agreed upon the strong hereditary component in social traits, the traditional opinion on the importance of heredity as an underlying cause of many human diseases was beginning to change because of the developments in bacteriology and public health after 1880. In this realm it appeared that nurture was more important than nature, that external factors from the environment rather than internal hereditary traits triggered organ malfunction that resulted in clinical disease.

The common understanding had been that the appearance of the same trait in parent and child resulted from heredity: Like begets like. The presence of a shared disorder at birth was felt to provide further evidence for the hereditary nature of the trait. This argument identified congenital tuberculosis (Cleveland, 1877) and syphilis (Witter, 1887) as hereditary disorders.

But even before the development of the microbial theory of infectious disease, there was uncertainty about the relevance of the hereditary paradigm in explaining the nature of human disease. When a particular disorder was observed to run in certain families, did that alone constitute proof that a hereditary predisposition was involved in the cause of the disease? As early as 1868 Carrington had summarized these concerns: "We seem to make an attempt to hide our ignorance when we talk in misty uncertainties such as 'diathesis or predispositions'" (Carrington, 1868–69). Longstreth agreed that "the determination of the hereditary nature of a disease is a matter of great difficulty" (Longstreth, 1882).

To answer this question on the importance of heredity as a cause of disease, physicians gradually recognized that the application of scientific methods to medicine was necessary. No longer would observation and personal experience be sufficient. Scientific knowledge, once viewed by the physician as unnecessary and esoteric, became "essential for everyone who would be considered a modern practitioner" (Ludmerer, 1985). Diagnosis and treatment were to be guided by science, not by impressions. As one physician claimed, "The rational practice of today is the applied science of yesterday" (Warner, 1991). Physicians were encouraged to improve their understanding of science because the application of these ideas was expected to improve the care provided to the patient at the bedside (L. King, 1991). J. H. Musser believed that scientific habits of thought would allow physicians to gather quantitative data about their patients and provide diagnoses that were "scientific, precise and positive." Critical thinking by physicians, not the application of empirical dogma based on personal experience, was essential for modern practice. "The first requirement of a diagnostician is not technical skill but a judicious mind, capable of weighing the volume of evidence" (ibid.).

In 1878 Louis Pasteur published his germ theory of disease. The application of this concept to animal, plant, and eventually human diseases over the next two decades appeared to demonstrate that much of pathology was due to these "offending microbes." But the relevance of this research for U.S. physicians at the time appeared to be problematic at best. When papers on bacteriology were read at medical society meetings, it was not unusual for prominent physicians to walk out, signaling their belief that this type of research had no clinical relevance (Rothstein, 1972). W. Belfield, criticizing the lack of knowledge of bacteriology by the nation's physicians, observed the "general state of ignorance in the medical community, the slow diffusion of scientific knowledge and the inability to evaluate evidence." In 1884 when H. D. Didama reported on developments in bacteriology at the annual meeting of the American Medical Association, he was disappointed to state that, in this instance, the possession of scientific knowledge did not benefit the patient: "Bacteriology with all its brilliant discoveries has furnished little help to what is of the greatest practical importance to physicians and patients, the art of healing" (L. King, 1991). J. A. Cutter summarized the opinion of many physicians in 1889: "We all aim to be therapeutists. Beautiful diagnoses obtained by scientific methods of examination are very nice, but if they do not cure the case, that is the treatment based on them, then we are at fault. . . . Unless we constantly hold therapeutics before our eyes, our profession will become nothing but one of scientific amusement" (Warner, 1986).

The application of scientific research to the diagnosis and eventual treatment of diphtheria finally provided the necessary proof that bacteriology would prove beneficial for physicians as they attempted to help patients. Diphtheria was known to be an epidemic disease, often killing several members of a family within a few days. Empiric treatments appeared to offer nothing to patients. Then, in 1884, the diphtheria bacillus was isolated in Koch's laboratory, and by 1890 it had been shown that the bacterium produced a toxin that spread throughout the body and produced the respiratory symptoms characteristic of the disease. Experimental animals injected with small quantities of the toxin developed protein antitoxin in their blood and were then resistant to subsequent infection with diphtheria. The antitoxin, isolated from the blood of these resistant animals and injected into other susceptible animals, made them resistant to the infection as well. If they already had the disease, the antitoxin appeared to slow the progression of the symptoms and improve the likelihood of survival. When large quantities of antitoxin were prepared in horses and used to treat human subjects during epidemics of diphtheria, mortality figures dropped by five- or sixfold (Rothstein, 1972). Instruction in bacteriology became a part of the regular medical school curriculum at Harvard in 1886 (Warner, 1986). By 1900 the success of the diphtheria antitoxin had convinced many physicians that bacteriology was essential for the accurate diagnosis and treatment of important human diseases (Rothstein, 1972).

This application of scientific knowledge began the radical transformation of U.S. medicine. Many other examples of infections caused by external agents were discovered during the last two decades of the nineteenth century. These external infectious factors appeared to explain the causes of many human ills. As a result, in the opinion of both the physician and the general public the relative importance of hereditary predisposition as an important cause of disease was greatly diminished (Beecher, 1960). This trend favoring the environmental causes of illness would gather momentum over the next thirty years, as will be examined in subsequent chapters.

Clinical observations were reinterpreted in the light of these new theories on the cause of human disease. The presence of a trait at birth was no longer necessarily attributed to heredity, for it could be the result of a congenital infection (Gaddy, 1883). Certainly by 1890 it was generally accepted that congenital syphilis, tuberculosis, smallpox, and glanders (an ulcerating skin disease) were infections transmitted from parent to child but acquired during the pregnancy (Pennington, 1877; Graham, 1885; Von Schaick, 1888; Rumbald, 1894; Jordan, 1898). "Infection was clearly the key here, not heredity" (Graham, 1885).

If heredity did not play a significant role in causing human disease, what function did it serve? After 1890 some physicians believed that heredity was the mechanism that preserved the basic structure of the body—the species type. An original form was the thing transmitted from parent to child. Nature did not perpetuate disease by heredity, as disease represented an imperfection that could result only from factors originating in the environment (Child, 1890). Most traits were healthy. The possibility of transmitting abnormal traits was believed to be limited because of the strong tendency of the natural system to revert to its original, more vigorous type (Hutchinson, 1892b). When disease did occur as a result of an environmental influence, the restoration of good health was dependent not only upon exterminating germs but also upon improving nutrition, vitality, and symmetry in the "noble bodies" (Hutchinson, 1892a).

In the 1890s arguments for and against the importance of heredity in the causality of disease raged within the medical community. Although in the areas of emotional and social traits its relevance was widely accepted, in the medical realm, the mere presence of the same disorder in a parent and child no longer necessarily implied hereditary transmission, for many infectious diseases followed the same pattern within families. Rumbald summarized one widely held view of nature and nurture when he pointed out that as specific causes of diseases had been identified, the list of supposedly hereditary diseases had diminished. In his opinion, "a predisposition to a disease may be inherited, but a disease is always the effect of an external cause . . . All influences which we inherited come from within and result only in normal activities . . . Every effect producing unfavorable results comes from without . . . Disease is brought about solely by environment" (Rumbald, 1894).

There was now general agreement that direct inheritance rarely if ever occurred. A predisposition could be transmitted which might result in disease if triggered by a specific, exciting external cause (Hartshorne, 1885; Richardson, 1896; Gordan, 1897). The bacillus of tuberculosis, for example, might infect individuals whose constitution made them unusually susceptible to such microbial invasions (Hutchinson, 1892a; Sisson, 1894). Disease was then a modified physiological process. When the limits of normal regulatory mechanisms had been exceeded, bodily functions became abnormal and physical disease began (Fitz, 1885). But the question remained as to which factors were important in altering the normal physiologic balance. Benedict spoke for many in the medical community when he suggested that "much of what is ascribed to heredity should be credited to infection, environment or chance" (Benedict, 1898).

CONCLUSION

In the 1890s the subject of heredity was an area of debate in both biology and medicine. There seemed to be much to understand, but no one agreed on the exact methods to be used to further that comprehension. As one physician described the situation, "The broad subject of heredity is complex and contradictory, because the factors that make up physical and mental characters are themselves complex and contradictory" (Crandall, 1897).

For the physicians of the day the major questions on heredity remained unresolved. First, was hereditary information worth having? Did it count for much in the daily management of human disease? Second, was hereditary information discernible? Or was the issue outside the realm of scientific investigation?

The examples in this chapter demonstrate the shift in emphasis from internal to external disease causality which began during the decades before 1900. The traditional assumption that much disease had a hereditary basis was seriously questioned as factors from the environment such as bacteria and fungi were identified as causes of disease. If this new model for disease was correct, then the notion of intrinsic hereditary disease became much less important. Disease predisposition could be inherited, but symptoms of disease resulted when an external trigger acted upon the susceptible individual. One physician was convinced that "when light is shed upon the etiology of all disease, the ghoulish cloak of heredity will be laid aside forever" (Rumbald, 1894).

The ill-defined concept of hereditary diathesis or predisposition to disease was discussed a great deal during these years, but it was a pathological model that was almost impossible to verify experimentally. Hence, it did not meet current standards for scientific fact. The improvement in general well-being brought about by public health measures to control infections also seemed to demonstrate the correctness of the external model for most human diseases.

Physicians in the United States were also acutely aware of the controversy in biologic circles over the mechanism of heredity in animals and plants, but they held a longstanding opinion that issues of human heredity were too complex for science to decipher. In 1878 it was observed that "many anomalies exist in heredity which the science of today is quite incompetent to explain" (Iles, 1878); in 1881 heredity was described as "an imponderable force" (Griswold, 1881); and in the 1890s the mechanisms of heredity continued to be a "mystery" (McMurrich, 1894; Stearns, 1897).

In the last decade of the nineteenth century many physicians had a rather pessimistic view of heredity as it applied to medical practice. The importance of heredity in causing human disease was questionable at best, and the technology to study it and understand its workings was limited. Perhaps in the future better models for human heredity would be developed. Sisson in 1894 expressed the hope that "some intrepid explorer would break into this secret storehouse of nature" (Sisson, 1894), and a year later Christison reported that "the nature of the fundamental law of heredity is unsettled and probably will be so for some time due to lack of sufficient facts" (Christison, 1895).

Biologists and physicians were equally dissatisfied with the current theories of heredity as applied to human beings. Their thinking in this area was in a state of turmoil. They were ready for some fresh ideas, ready for a change. The rediscovery of Mendel's work in 1900 would unleash a torrent of facts on heredity and revive the controversy on the role of heredity as an important mechanism of human disease.

THREE

MEDICINE AND THE NEW SCIENCE OF GENETICS
1900 to 1910

It will be epoch-making, as important as the atomic theory. —*Popular Science Monthly*

It may seem . . . to be a too entirely theoretical subject . . . for busy medical men. —*Philadelphia Medical Journal*

As the twentieth century began, the spectrum of opinion on the importance of heredity in biology and medicine was broad indeed.* The stage had been set in biology for an appreciation of new hereditary models because of significant changes in thought on the role of heredity within the general areas of embryology and evolution. Experimental work in the previous decades had demonstrated that the transmission of specific characters from one generation to the next was a separate issue from that of the growth and development of the entire organism. Experimental breeding had shed light on the sources of variation necessary for the natural selection theory of evolution. By the end of the nineteenth century, experimental breeding had become a legitimate scientific field in its own right. The time was ripe to redefine the significance of embryology and heredity in the general field of growth and reproduction (Wilkie, 1962; Bowler, 1989).

Experimental work in heredity flourished in the United States after 1900, reflecting the general shift in biology from a descriptive, morphologic natural history discipline to a hard science with facts derived from experiments. Biology in the major universities at this time strove for precision and recognition just as did chemistry or physics (Cravens, 1978). When European plant breeders rediscovered Mendel's work in 1900, it was rapidly accepted by U.S. biologists, who had the technical and intellectual background to appreciate its relevance to their own breeding studies (Paul and Kimmelman, 1988). "The spec-

---

*The epigraphs are drawn from Cook (1903) and Wright (1902).

ulative writings of Darwin and Weismann on the mechanisms of inheritance had led the geneticists of the last quarter of the century quite deeply into a muddle which was only gradually being resolved by microscopic observations on chromosomes; and one would imagine that Mendel's thinking would seem strikingly clear to a geneticist of the year 1900" (Lanham, 1968).

The evidence reveals, however, that this decade of discovery was characterized by controversy in both biology and medicine as to not only the validity but also the relevance of Mendel's work for the understanding of heredity in plants, animals, and humans.

THE REDISCOVERY OF MENDEL'S WORK

The traditional story of the rediscovery of Mendel's work involves three German plant breeders working independently around the turn of the century who found evidence for discontinuous inheritance of particular traits in several species. DeVries presented his theory of inheritance and the purity of the germ cells, which coincided strikingly with the model proposed by Mendel in 1865 (1900). Within months, Correns (1900) and Von Tschermak (1900) reported similar results. All three found that the hereditary principles outlined in Mendel's (1865) paper provided precise explanations for the transmission of discrete characters from one generation to the next. The English biologist Bateson had also found evidence for discontinuous variation in his plant breeding experiments during the 1890s. He learned of these three reports in 1900, immediately appreciated the usefulness of Mendel's ideas for hereditary work in general, and presented them to the English-speaking world late that year (1900–1901). By early 1902 he was able to demonstrate the inheritance of unit characters in animals as well as plants (1902).

Many biologists in the United States, where plant and animal breeding was a new and active field of research, quickly recognized the usefulness of Mendel's theories for their own work in heredity. Over the next ten years, heredity developed as an independent branch of science with its own methods, terminology, professional organizations, and publications. Independent work in breeding and cytology had reached similar conclusions on the mechanisms of heredity in plants and animals, and the findings appeared to be useful for daily scientific work (Paul and Kimmelman, 1988; Bowler, 1989). "Nothing is probably more favorable to rapid advances in science than the discovery that two independent fields of investigation can be so connected that their results are mutually both stimulative and corrective" (Wilkie, 1962).

The enthusiastic reception of Mendel's work by U.S. scientists generated a flurry of reports—both experimental and theoretic—that defined classical genetics by 1910. A summary of Mendel's laws for inheritance was first reported by Charles Davenport, a biologist at the University of Chicago who had been involved in plant and animal breeding for more than ten years. He outlined notions of dominant and recessive characters, the independent segregation of these characters at meiosis, and the idea that only one "antagonistic peculiarity" was carried by each germ cell (Davenport, 1901). Another early advocate of Mendelism was Wlliam Castle, a breeder of mammals at Harvard University who proposed that a compound unit of two or more characters might exist which was not separated during meiosis. These "coupled characters" could then be transmitted together in the germ cell from one generation to the next (Castle, 1902–03, 1903d). The inheritance of specific traits in animals was soon interpreted in Mendelian terms. Coat color in mice, rabbits, and guinea pigs was inherited in conformity with dominant and recessive rules (Castle, 1903c; Castle and Allen, 1903), and hair length in rabbits and guinea pigs appeared to behave as a recessive trait (Castle, 1903b).

Observations by Wilcox in the field of cytology as early as 1901 suggested that the study of chromosomes would "elucidate the mechanical means of heredity." He agreed with Weismann that the hereditary tendencies resided within the particles that comprised the chromosomes (Wilcox, 1901). Cannon and Wilson independently presented observations on the segregation of maternal and paternal chromosomes during meiosis, which resulted in germ cells, half with paternal and half with maternal chromosomes. This symmetric distribution of characters in the germ cells was believed to provide "a physical basis for the association of dominant and recessive characters in the cross breed . . . exactly as the Mendelian principle requires" (Cannon, 1902; Wilson, 1902).

Over the next two years, Montgomery and Sutton were able to state confidently that the segregation of characters and chromosomes was in "exact correspondence." The paired arrangement of the chromosomes implied a dual basis for the inheritance of each character; and each chromosome had to carry several characters. The segregation of these coupled characters would be coincident, although their expression might be independent (some dominant and others recessive) (Sutton, 1903; Montgomery, 1904). The segregation of chromosomes thus appeared to represent the physical basis for the transmission of the unit characters as outlined by Mendel.

One specific example of the interaction between cytology and heredity involved the mechanism for sex determination. In early 1903 Castle proposed a Mendelian mechanism for the inheritance of sex based upon the segregation of unit characters (Castle, 1903a). McClung and Stevens, however, suggested a different mechanism after noting the existence of a different number of chromosomes in male and female cells of certain insects. Their breeding studies indicated that the presence of an "accessory chromosome" was involved in the determination of sex (McClung, 1901; Stevens, 1905–06).

By 1910 the study of heredity had been redefined as the new science of genetics, a distinct area for experimental investigation and rapidly becoming a preeminent U.S. science (Cravens, 1978). T. H. Morgan of Columbia University reviewed these dramatic developments in a major paper in that year. He believed that chromosomes in cells from different body tissues were identical. Genetic continuity meant the segregation of identical chromosomes into the daughter cells during mitosis. Because different chromosomes appeared to serve different functions, normal embryonic development required the interaction of products from the entire complement of chromosomes. Each chromosome, he wrote, probably carried many characters; the segregation of each chromosome pair paralleled the segregation of unit characters during meiosis. In this respect genetics and cytology united to explain a mechanism for the hereditary transmission of traits from one generation to the next (Morgan, 1910).

The accumulation of experimental data on hereditary mechanisms convinced many biologists that each trait was represented by a determiner in the chromosomes, which then segregated according to Mendel's ratios. Not all workers in this field were convinced, however, that the unified genetics-cytology model provided an all-encompassing scheme for heredity. The decade was, in fact, a time of fierce controversy in the biologic literature (Cravens, 1978; Bowler, 1989). Prominent biologists in Europe and the United States rejected the new model, either in part or in total. Montgomery argued that the accessory chromosome might "follow sex or be associated with other differences that determine sex" in animals, rather than being its sole cause (Montgomery, 1910). Guyer believed that the cytoplasm was most important for normal development. Observing that closely related species often had widely divergent numbers of chromosomes, he argued that this divergence was out of all proportion to the actual differences between the lines. The chromosomes might carry "superficial qualities, " while the basic body structure was governed by factors in the cytoplasm. Heredity from previous generations merely brought "shaping and controlling factors" into place (Guyer, 1907, 1909). Conklin and Bateson made similar observations: changes in chromosome number or structure appeared to have little to do with the development of specific characters. Hence, they doubted the importance of chromosomes as determinants for embryogenesis (Bateson, 1907; Conklin, 1908).

Despite these controversies, genetics as a new experimental science developed rapidly during the first decade of the new century. Biologists could now do real quantifiable science, not just observe natural phenomena. The increase in professional activity in genetics seemed to indicate to both scientists and the public at large that hereditary factors were important in the development of animals and plants. In the nature-versus-nurture discord, nature seemed to have become more prominent. Not all scientists had adopted a hereditarian attitude, but heredity certainly "loomed large" in the rapidly changing new science of genetics (Cravens, 1978).

HUMAN CHARACTERS AND MENDELISM

The successful application of Mendel's theory of inheritance to traits in plants and animals set the stage for parallel studies in humans. The Bussey Institute at Harvard University was the location of much early work in human genetics. Its director, William Castle, had studied zoology at Harvard with Davenport and had received the Ph.D. in 1895. Appointed to the Harvard faculty in 1899, he had begun breeding studies on several different species of animals. After publishing studies on mice, rats, rabbits, guinea pigs, and other domestic animals, he rapidly became recognized as the most prominent mammalian breeding scientist in the country (Morse, 1985).

A graduate student, W. C. Farabee, examined a family in which several albino children had been born to a man who was normally pigmented himself. He had been married to two normally pigmented women but had an albino father. The family lived in an isolated rural area where several other albino families were known. Farabee and Castle suggested that the inheritance of albinism in this example could be interpreted as a recessive trait, just as it was in other mammalian species. They proposed that the father and each of his wives carried the albino trait in a recessive state. According to Mendelian predictions, one in four of the offspring should be albino. The family actually had four of fifteen affected, a close approximation to the expected ratio (Farabee, 1903; Castle, 1903e).

The research for Farabee's doctoral dissertation in 1903 involved an analysis of the heredity of a human hand malformation, brachydactyly. Members of five generations in one family were affected. Males and females were equally involved; about half the offspring had the anomaly. He concluded that "the present case demonstrates that the law [of Mendel] operates in man as in plants and lower animals. The abnormality is shown here to be a dominant character" (Farabee, 1905).*

At the same time, another Bussey colleague studied the inheritance of human polydactyly, a second malformation of the hand. C. W. Prentiss cited the sudden appearance of such reversionary abnormalities or "rogues" as evidence that hereditary characters could be transmitted in a latent state from one generation to the next. When active, the character appeared as a new feature of the individual's body structure. Although he believed that the law of Mendel applied to the mechanisms of human hereditary transmission, he thought that another model would be necessary to explain the origin of new variations within the germplasm (Prentiss, 1902–3).

A second early center for research in human genetics developed at the Cold Spring Harbor Station for Experimental Evolution, on Long Island, New York. Charles B. Davenport, named director in 1904, had originally been trained as a civil engineer and did railroad surveying after graduating from the Polytechnic Institute in Brooklyn, New York. He became interested in zoology and completed the Ph.D. at Harvard in 1892, where he taught until 1899, when he was recruited for the faculty of the new University of Chicago. During summer vacations, he directed the Biology Laboratory of the Brooklyn Institute of Arts and Sciences, which was held at Cold Spring Harbor. He became a successful breeder of plants and animals, and corresponded with leading workers in the field in both the United States and Europe. After 1901 he served as the U.S. editor for the British heredity journal *Biometrika* (Rosenberg, 1961; Gillispie, 1976).

Davenport initiated a series of studies on human characteristics

---

*Despite these promising beginnings, Farabee did no further work in human genetics. Instead, his professional career was devoted to anthropology (Farabee, 1912, 1918).

that appeared to be inherited, beginning with eye color. Blue appeared to be recessive to brown, and brown was dominant to blue. The pedigrees he collected were interpreted in simple Mendelian terms (Davenport and Davenport, 1907). Hair form also seemed to follow Mendelian ratios. Straight hair resulted when an individual inherited the straight factor from both parents, curly when that factor was present in double dose, and wavy when the individual received the straight factor from one parent and the curly factor from the other (Davenport, 1907–8; Davenport and Davenport, 1908). Hair color was explained by the segregation of two major color traits: red and brown. Davenport concluded that the inheritance patterns did not fit the results predicted by blending models because most offspring did not have an intermediate hair color relative to their parents'. The Mendelian model appeared to provide a more accurate description of the actual results (Davenport and Davenport, 1909). He also found evidence that human albinism followed the recessive pattern of inheritance. The transmission of blond and brunette skin pigment generally demonstrated the expected ratios, with brunette being dominant and blond recessive (Davenport and Davenport, 1910).

About the same time, Spillman reported a family in which the parents were first cousins. Of seven children, two were deaf. He interpreted this pattern as characteristic of a recessive trait. The parents were heterozygous, and approximately 25 percent of the progeny were homozygous recessive, and therefore demonstrated the physical impairment (Spillman, 1905).

Analyzing the inheritance of other human conditions was not so straightforward, Davenport found. To study the segregation of black and white skin pigmentation, he attempted to quantitate skin pigment using a color chart but concluded that perhaps a "myriad of unit characters" was responsible for the continuous variation in pigmentation which he observed (Davenport and Davenport, 1910). That the same trait in different families did not necessarily follow the same pattern of inheritance added to the analysts' problems. In one example of polydactyly the trait followed a dominant pattern. But in another family the data did not appear to fit Mendelian ratios at all. Deafness in several families was evident in successive generations, but Davenport could not interpret the results in terms of either Galton's or Mendel's laws of inheritance (Davenport 1904). In similar studies, Holmes examined the inheritance of eye and hair color in seventy-one families and concluded that these traits existed in a continuum, not as discrete entities, and that their patterns of inheritance did not fit the ratios predicted by Mendel's laws. He concluded that such inheritance was to a certain extent of the blended type (Holmes and Loomis, 1909–10).

CONTROVERSY IN MEDICINE: DIATHESIS VERSUS GENETICS

From the muddle of excitement and controversy over Mendel's laws in biology, the first tentative report on new developments in heredity appeared in the U.S. medical literature. M. F. Guyer published "The Germ Cell and the Results of Mendel" in the *Cincinnati Lancet Clinic* for April 1903 (Guyer, 1903). He was a zoologist at the University of Cincinnati who had studied spermatogenesis in hybrid pigeons (Guyer, 1900, 1902a, b). His article outlined the nature of dominant and recessive characters, and the proportion of each expected in matings of different parental types. Without discussing the possible application of these findings to human traits, Guyer concluded that "the facts . . . seem to point precisely to such a state of affairs in the germ cells as the law would necessitate."

In the following years of the new decade, other biologists involved in hereditary research were invited to present their findings at medical society meetings, a primary educational source for the communication of new ideas on heredity to the practicing medical community. The prestigious Harvey Lecture at the New York Academy of Medicine in 1907 was given by E. B. Wilson from Columbia University. Wilson said that the unit character type of heredity was quite different from the theories of blending inheritance. Such characters could be brought together during fertilization, separated during meiosis, and recombined in subsequent generations, following definite proportions. The physical representation of these hereditary units appeared to be the chromosomes in the cell nucleus. In his opinion, "chromosomes stood for or are the physical basis of corresponding characters," providing a "fundamental explanation of Mendelian inheritance." He suggested that "in inheritance we are not dealing with vague questions, but clear, concise mathematical problems" (Wilson, 1907a, b, 1908).

Human genetics was the specific topic of the Harvey Lecture presented by C. B. Davenport two years later. His review of the recent research had convinced him that Mendelian heredity operated in humans just as it did in plants and other animals: "The study of human heredity is the study of the laws of the development into somatic unit characters of the determiners that are brought together in the union of the germplasms." But he found the available data on the inheritance of specific human traits "meager and inadequate" and suggested that practicing physicians could make significant contributions to the study of human genetics by reporting families with conditions that recurred in successive generations. He was convinced that such knowledge would have immense practical value, for providing hereditary information to the public would result in "a fair proportion of educated reasonable people making a selection of their consorts with a fair regard for the probable nature of the offspring" (Davenport, 1910c). He believed that cooperation between scientist, physician, and the public would result in the better children necessary for progress in U.S. society.

The Chicago Medical Society also sponsored a conference on heredity in 1909, at which the keynote speaker was W. E. Castle, who reviewed the laws of Mendel for the independent segregation of characters during meiosis. Both parents appeared to share equally in hereditary transmission. In his opinion, this transmission involved the passage of "enzyme-like materials which initiate certain metabolic processes in suitable medium represented by the food materials of the egg" (Castle, 1909).

From the vantage point of the practicing physicians, all of these discussions probably seemed esoteric. None of the researchers was able to provide one example of an important human trait that was inherited in a fashion that could be explained by the application of Mendel's laws.

The mere awareness of these new ideas on heredity certainly did not induce physicians to radically alter the practical rules for heredity which had served them reasonably well over the years. As Mendel's laws were discussed in the medical literature, two distinct schools of thought on the nature of heredity developed in the United States. For the first and larger school, the traditional notion of heredity as diathesis or predisposition to disease remained the dominant paradigm throughout the first decade of the new century. Heredity was defined as the direct transmission of a character from parent to child (Horne, 1902; Riley, 1902). The actual expression of disease almost always required an external trigger from the environment to act upon the altered organ system (Sager, 1900). In general, specific diseases were not inherited. Rather, a weakness in one system might be manifested by different symptoms of disease in different members of the same family (Kiernan, 1902b; Chase, 1907). Heredity was now viewed not as an event but as a process that involved something transmitted but modified by the

Table 3.1. Hereditary Predispositions Reported, 1900–1910

Trait	Reference
Direct heredity	
Ankylosis	Walker, 1901
Muscular dystrophy	Marvin, 1902–3
Epilepsy	Meyers, 1904
	Spratling, 1910
Hypospadius	Strong, 1906
Chorea	Hamilton, 1908
	Tilney, 1908
Special heredity	
Ichthyosis	Bromwell, 1902–3b
Hemophilia	Hicks, 1903
Collateral heredity	
Amaurotic family idiocy	Hymanson, 1902
Epilepsy	Meyers, 1904
Epidermolysis bullosa	Engman and Mook, 1906
Ménière disease	Walker, 1910

environment during embryonic development. "Heredity obeys no absolute laws, but is governed by a struggle between contending forces" (Kiernan, 1902a).

The four hereditary patterns described by Ribot (1875) continued to prove useful as clinicians examined the role of heredity in human disease (Whetstone, 1902; Barr, 1904; Hilton, 1905). As late as 1909 Loeb used them to describe the hereditary nature of 496 cases of blindness (Loeb, 1909), and throughout the decade Ribot's types were identified with the transmission of specific disease predispositions (table 3.1). The strongest evidence for heredity remained the direct transmission of the trait from parent to child: "Like begets like." The peculiar pattern of transmission in which unaffected females appeared to transmit a character to their sons, as in hemophilia, was also viewed as a strong indication that heredity was at work (Hicks, 1903).

The second school of thought included a growing number of physicians who were dissatisfied with the traditional understanding of hereditary transmission. Whenever several people in a family had the same symptoms, a hereditary predisposition was inferred, but the genuineness of heredity was less obvious. Benedict suggested that several criteria should be met if heredity was actually involved: (1) The character should be transmitted to all, or nearly all, of the progeny; (2) the character should be transmitted to several generations and to collateral lines in the family; and (3) the trait should persist despite environmental factors opposed to its development (Benedict, 1902).

The mere presence of the condition in several generations of a family was inadequate evidence for heredity as a causal factor. For example, the occurrence of the neurologic degenerative disease amaurotic family idiocy in siblings and cousins could represent an infection, perhaps passed in the mother's breast milk (Hymanson, 1902). Cataracts were present in four generations of another family, but did this necessarily prove that heredity was the cause? (Wood, 1906). The belief that a predisposition to disease was transmitted was also being questioned. One author believed that this whole notion was "inexplicable" (Sinkler, 1906).

The failure of existing theory to suggest a mechanism for the transmission of disease diathesis was frustrating. Bunting and Hanes, two physicians from Johns Hopkins Hospital, wondered what exactly was involved in this notion of heredity. Bunting described several families with cystic kidney and liver in which affected parents had produced both healthy and affected children; the character in the parent thus was not always transmitted to the progeny. He regretted that any further understanding of this factor was impossible (Bunting, 1906). Hanes also felt stymied by his inability to explain the familial nature of "hereditary" telangiectasia, a condition characterized by unusual bleeding from the capillaries. He noted that "to define with greater peculiarity this 'hereditary tendency' would be a pleasure, " but "to find the specific fault which underlies the disease is a problem for the future" (Hanes, 1909). An explanation for the apparent transmission of other traits was simply impossible with the existing theories of inheritance (Larrabee, 1906). As late as 1910, one physician regretted that the "modus operandi" of heredity remained an unresolved problem (Spratling, 1910).

POSSIBILITIES FOR CHANGE: THE NEW GENETICS AND MEDICINE

The controversy on the role of new discoveries in heredity and cytology was known to many physicians during this time. Although not active participants in biological research, they sensed that something new was in the wind. As they became more familiar with these new ideas on genetics, they also began to question whether inheritance in humans, especially as it related to disease, followed similar laws of nature.

A Canadian with close ties to many prominent U.S. physicians, J.

George Adami was an early interpreter of the role of heredity in medicine. A student at Cambridge during the early 1880s, he had been caught up in the scientific excitement there and had come to perceive medicine as a branch of biological science. He formed a close friendship with William Bateson, one of his classmates at Christ's College. After further medical training in England and Germany, he was appointed to the teaching faculty at Cambridge, where he functioned as a research pathologist and studied cardiovascular physiology and infectious diseases.

In 1892 Adami was named professor of pathology at McGill University in Canada, where he established a clinical research unit. He visited the United States many times, became friends with William Osler and William Welch at the Johns Hopkins Hospital, and was eventually honored for the quality of his work by election as president of the Association of American Physicians (Adami, 1930).

Throughout his professional career, Adami examined the important causative role that heredity might play in human diseases. He also recognized the confused state of affairs in biology and medicine regarding different theories of heredity. In 1900 he outlined a molecular model for inheritance at the Brooklyn Medical Club, unaware of the recent reports on Mendel's work. He agreed with Weismann that the egg and sperm carried equal numbers of hereditary units. He believed that inheritance occurred at the union of egg and sperm cell nuclei during fertilization, and that variation was the result of this "fortuitous comingling of ids." He proposed that the germplasm had a definite molecular structure that was quite stable but could be altered. The complex "proteid" material of the cell nucleus appeared to be the germplasm, structured as a series of rings to which were attached side chains. Egg and sperm would contribute different side chains, thus producing variation in the offspring. Environmental factors could possibly cause the loss of certain side chain units. If such altered germplasm was then transmitted to the subsequent generation, the model would explain the inheritance of acquired conditions (Adami, 1901a, b). Adami assumed a strict hereditarian model: the physical makeup of each individual was primarily the result of the structure of the germplasm.

Some physicians at this time began to consider the possibility that the same biologic laws for heredity which applied to plants and animals also applied to humans. One noted that Mendel's laws could predict breeding results with "almost mathematical accuracy" (Morrow, 1907). But the understanding of such technical issues was quite difficult, given the limited scientific training of most practicing physicians. Irwell observed that "the average medical man does not even understand the terminology used by those workers whose efforts are directed toward the unraveling of biologic phenomena" (Irwell, 1902b).

The various patterns in which predisposition appeared in successive generations seemed to suggest that heredity involved more than the direct transmission of a trait from parent to child (Talbot, 1904). If these new developments could provide better answers than the existing heredity models, the investigation of human characters should offer insight into the general mechanism for heredity, as the work published by Charles Dana exemplifies. Dana was professor of diseases of the nervous system at Cornell University Medical College in New York and served as president of the American Neurologic Association in 1892 and again from 1906 to 1907. Although he was a leading researcher on the correlation between changes in the structure of the nervous system and clinical symptoms of disease (Kaufman, Galishoff, and Savitt, 1984), his published work in this decade reveals no familiarity with the new developments in human heredity. Studies of the muscle disorder myoclonus multiplex and pernicious anemia in several generations included no comments on the role that heredity might play in the etiology of such diseases (Dana, 1903, 1907). Dana later reported that he had become aware of the "new" heredity in 1906 or 1907 and thought that it might explain a great deal about the development of human neurologic disease (Dana, 1910). He commented that Huntington chorea had always appealed to him as an excellent disease in which to work out and apply the Mendelian theory, and he hoped that research workers in this field would consider its hereditary feature (Huntington, 1910). Unfortunately, he apparently did not understand enough about heredity to analyze the family data himself.

In a similar vein, Buchanan's study of the inheritance of cleft palate led him to suggest that medical professionals "will come to take up Mendel's law of heredity, study it and apply it to these cases." With this approach, he believed, one could calculate the likelihood that other children in a family would be affected with the same disorder as their parent (Calvin, 1908). But the ability to analyze human inheritance with the Mendelian theory eluded these physicians. As Calvin observed, "Heredity is one of the subjects that medical men have not studied enough." He had no idea who Mendel was, however, and recommended Ribot's work on the subject (Calvin, 1908).

Adami prepared a thirty-three-page chapter on inheritance and disease for a popular multivolume text in 1907 entitled *Modern Medi-*

cine, for which Osler, as editor, had solicited chapters on clinical medicine from prominent physicians in the United States, Canada, and Europe. Here, Adami attempted to unify his molecular model for germplasm with the new Mendelian theory. He proposed that specific characters were represented by specific side chains attached to a central core molecule within the germplasm. Meiosis involved the separation of maternal and paternal traits, and fertilization then united genetic material from both parents in equal amount. The segregation of dominant and recessive characters in meiosis, a result of the variable potency of particular side chains, appeared to fit the proportions "demanded by Mendel's law" (Adami, 1907). Although the theory sounded plausible, once again its application to specific human disease traits proved formidable.

Physicians thus recognized that there was something to these new ideas, but exactly what was it? McKee suggested that altered cellular metabolism might account for the degeneration of the neurons observed in amaurotic family idiocy, for example. For the actual seat of the defect, he proposed that "it may be to the sperm or to the ovum that we must look, and possibly to the chromosomes of these cells" (McKee, 1905). The behavior of chromosomes also could explain the existence of particular hereditary predispositions in only one sex. Clowe reported observations on the segregation of accessory chromosomes in experimental animals. The existence of such chromosomes in humans might explain why certain traits appeared to be transmitted only through the female line of a family (Clowe, 1909).

The segregation of the chromosomes also suggested a mechanism for the frequent occurrence of hereditary traits within inbred families. Cytogenetic observations of meiosis documented the loss of half of the chromosomes from precursors of egg and sperm. If the egg and sperm from related male and female contained much of the same hereditary material, there would be an increased likelihood of the fertilized egg containing defective germplasm from both sides of the family (Dean, 1903). This intensification of qualities through inbreeding was the logical consequence of Mendel's new theory of inheritance (Libby, 1908).

Particulate inheritance seemed to make sense in such cases. But the sudden appearance of new characters in individuals also required an explanation. It was proposed that the mutation theory of DeVries might explain changes within the germplasm, which could then be transmitted to subsequent generations (White, 1909).

By the end of the decade, leaders in medicine were calling for more attention to the details of human heredity. Before a large gathering of practitioners Dana discussed the new developments in the field of heredity. "It is such pronouncements as these, coming from authoritative sources and backed up by an enormous mass of statistical and laboratory studies, that makes it seem important for medical men to take a little more interest and inform themselves a little more of what is being done day to day by the students of the problems of heredity" (Dana, 1910).

HEREDITY AS A CAUSE OF DISEASE

The primary reason for physician interest in these new developments within genetics was their potential value for explaining mechanisms of human disease. With the growing importance of science within U.S. medicine, there was hope that the application of this knowledge might result in better treatment for the patients afflicted with diseases having a hereditary predisposition.

But by the end of the decade there was pessimism in both medical and biologic circles on the relevance of heredity to human disease. The physicians were impressed by the differences between individual human beings: if the determinants of heredity really involved a physical arrangement of molecules within the germplasm, an enormous number of different forms must be present in the human population. It seemed improbable that such a chemical mechanism would ever fit the requirements for human heredity (Wright, 1909). Dana believed that the law of Mendel might apply to some individual characters, but as humans were so complex, one character often "would be submerged by the dominance of others" (Dana, 1910). Adami shared the opinion of leading British workers such as Bateson. Despite the molecular model for the germplasm he had devised, that is, he viewed human heredity as so complex that Mendel's laws could describe the issue in only general terms. He felt that it did not explain much, "anymore than the law of gravitation" (Adami, 1907). Woods summarized the opinion of many physicians by noting that "Mendel's laws, so important to the horticulturist and to the breeder of superficial traits of domesticated animals has not been shown to have any bearing on human heredity, at least as concerns important characteristics" (Woods, 1908). Clowe expressed the same opinion: "All the evidence goes to show that it [the law of Mendel] has no influence on man" (Clowe, 1909).

A discussion of the role of heredity in the recurrence risk for mental illness presented by Dana in 1910 epitomizes the state of genetics in medicine at this time. After discussing in detail the facets of Mendel's laws for heredity, Dana applied the classification system proposed by Ribot in 1875 to analyze the risk for hereditary transmission of mental illness. An individual whose family exhibited direct heredity of mental illness—that is, where a parent and child had both been mentally ill—took "the greatest risk in marrying." The presence of disease in only one collateral line indicated a slight risk for recurrence, but when it was evident in collateral lines from both sides of the family, the danger to a marriage was felt to be "moderately great." If an abnormal character had been present in the distant past but had not reappeared in three generations, Dana was confident that it would not recur in subsequent offspring from that line (Dana, 1910).

Opinions such as these from the recognized leaders in U.S. medicine suppressed the nascent interest in human heredity during this first decade after the rediscovery of Mendel's work. Physicians were left to ponder what all the controversy might mean for them in the daily practice of medicine.

The newness of the science of genetics was recognized by all the workers in the field. When P. C. Knapp discussed the potential role for heredity in human disease at a meeting of the American Neurological Association, he observed that "the application of Mendel's theory to human heredity has barely begun, and the complexity of the problem is so great that it is a question whether its potential bearings can be determined for many years to come" (Knapp, 1907).

The new genetics had therefore not yet defined its role in medicine. Practitioners continued to use the traditional "like begets like" mechanism for heredity, which helped to organize their family data but appeared to explain very little. Friedreich ataxia, for example, examined in several hundred families, was only occasionally observed in both parent and child. Much more commonly, several siblings were affected, while parents and other relatives were not. It was therefore a familial disorder, but not hereditary (Sinkler, 1906; Hunt, 1910). Robins felt that the "question of heredity is left to our imagination" in such cases (Robins, 1906–7). The presence of amaurotic family idiocy in several children of a family whose other relatives were unaffected did not make sense either, if this were a hereditary condition. The traditional concept of heredity would predict that most of the children would be afflicted. In many of these families, in fact, less than half were diseased (Frank, 1906). The application of the theory of heredity to actual clinical cases seemed problematic at best.

In the cases of Friedreich ataxia and amaurotic family idiocy re-

Table 3.2.   Heredity in Friedreich Ataxia and Amaurotic Idiocy,
1900–1910

Type of Occurrence	Reference[1]
Friedreich ataxia	
Direct	Barker, 1903
2 siblings (4 cases); single case	
(4 cases)	Collins, 1903
Single case	Frazier, 1903
2 siblings; collateral: 2 aunts	Hoppe, 1903
2 siblings	Van Wort, 1904
Single case	Dodge, 1906
Single case	Weber, 1906
2 siblings	Robins, 1906–7
Single case	Moore, 1908
Single case	Neff, 1909
3 siblings	Grannan, 1910
2 siblings	E. L. Hunt, 1910
2 siblings (3 cases); single case	E. H. Hunt, 1910
Single case	Spiller, 1910a
Amaurotic family idiocy	
2 siblings	Sachs, 1903
3 siblings	Frank, 1906
Single case	Buchanan, 1907
Single case	Cohen and Dixon, 1907
3 siblings	Brooks, 1909
2 siblings	Davis, 1909
2 siblings	Wandless, 1909
Single case	Howard, 1910
3 siblings	Sachs and Strauss, 1910
2 siblings	Strauss, 1910

[1] None of these studies provided pedigrees.

ported during this decade, in no instance was the pedigree used to
outline the presence of disease in different generations (table 3.2). One
example of direct heredity of ataxia was observed, but this did not alter
the opinion of most U.S. physicians that this type of familial pattern of
disease was not suggestive of hereditary predisposition.

Physicians regularly used the traditional definition of heredity to
classify human disorders. Direct parent-to-child transmission of the
characteristic was accepted as important evidence that a hereditary
predisposition was involved in the cause of the disorder; its absence
usually implied that hereditary factors did not play an important causative role (table 3.3).

Table 3.3.   Hereditary Nature of Human Diseases, 1900–1910

Disease	Heredity as Cause?	Reference
Direct heredity		
Fascioscapulohumeral		
muscular dystrophy	Yes	Marvin, 1902–3
Chorea	Yes	Cottral, 1905
Ichthyosis	Yes	Shoemaker, 1907
	Yes	Schwartz, 1909
Polycystic kidney	Yes	Neate, 1909
Syndactyly	Yes	Griffith, 1910
Spastic spinal paralysis	Yes	Spiller, 1910b
Sex-linked heredity		
Hemophilia	Yes	Hicks, 1903
Collateral heredity		
Cleft lip and palate	Yes	Brophy, 1901
Single case		
Ichthyosis	No	Wymer, 1908–9
Polycystic kidney	Yes	Bovee, 1909
Myatonia congenita	No	Laffer, 1909
Cataract	Yes	Robin, 1909–10
Spastic paraplegia	Yes	Lawrence, 1910

The pedigree, or family tree, continued to serve as a powerful representation that heredity as it was traditionally understood was involved in causing human disease. Almost without exception, the ability to diagram direct parent-to-child inheritance was accepted as evidence that hereditary factors played an important role in causing diseases of many different organ systems. In nineteen examples of direct inheritance documented by pedigree during this era, only optic atrophy and hemophilia appeared to fit the special sex-linked model for inheritance, although one family with hemophilia did not fit any known model for hereditary transmission (table 3.4).

In a widely distributed article, Munson recommended the use of the pedigree for reporting family histories, employing the modern symbols of circles for females and squares for males (Munson, 1910). The pedigree was used by an increasing number of physicians and biologists to represent their interpretation of the importance of heredity in cases of family disease.

Hereditary disease predispositions derived from the medical literature of this time range from alcoholism to hip dislocation to rickets (table 3.5). The major question for practitioners was whether the avail-

Table 3.4. Heredity Studies Providing Pedigrees, 1900–1910

Trait	Reference
Direct heredity	
Ankyloses	Walker, 1901
Spastic spinal paralysis	Spiller, 1902
Ichthyosis palmae et plantae	Bromwell, 1902–3a
Myopia	Frank, 1903–4
Angioneurotic edema	Fairbanks, 1904
Optic atrophy	Knapp, 1904
Ectopia lentis	Lewis, 1904
Dupuytren contracture	Mandoff, 1904
Spinal muscle atrophy	Collins, 1905
Cerebellar ataxia	Neff, 1905
Hemeralopia	Bordley, 1908
Chorea	Hamilton, 1908
	Tilney, 1908
Neurofibromatosis	Harbitz, 1909
Myotonia congenita	Sedgwick, 1910
Sex-linked heredity	
Hemophilia	Hicks, 1903
Optic atrophy	Mix, 1903
Unclear heredity	
Hemophilia	Larrabee, 1906

able understanding of heredity could assist them in interpretating disease mechanisms and their potential for transmission from one generation to the next. F. A. Woods was pessimistic about Mendelism as a solution, observing that "Mendelian traits usually seem to be simple and unimportant." He believed that human disease was often the result of many different factors and could rarely be explained in terms of dominant or recessive segregation (Woods, 1910). The general consensus in medicine continued to be that specific human disease was not hereditary in most instances, only predispositions were (Graves, 1910). The general human body form was certainly a common factor passed from parent to child, but the changes in function that represent disease were increasingly identified as the result of environmental influences instead of internal hereditary factors (Blackwell, 1903). The presence of a disease state in several members of a family was often viewed as the result of identical environment rather than identical heredity (Robins, 1906–7). Rogers observed that "the Mendelian laws are liable to be overworked to explain manifestations that are due entirely to environment" (Rogers, 1909–10). As practical people, physicians were increas-

Table 3.5   Hereditary Diseases Reported in 1907

Acromegaly	Leukemia
Alcoholism	Myxoedema
Alopecia	Ichthyosis
Arteriosclerosis	Insanity
Asthma	Morphine abuse
Diabetes	Nephritis
Epidermolysis bullosa	Obesity
Epilepsy	Psoriasis
Glaucoma	Retinitis pigmentosa
Gout	Rheumatism
Grave disease	Rickets
Hemophilia	Scoliosis
Hip dislocation	Tuberculosis

Source: Data from Morrow, 1907.

ingly inclined to look for toxins and microbes as the agents that upset the normal physiologic balance, a quest that was often successful. Because it worked, it seemed to be right.

At the same time, the opposite trend was evident within the biological community, where the belief in the relative importance of heredity as a mechanism for variation was increasing as research workers studied the regular transmission of characters from one generation of experimental plants and animals to the next. The disparity between the hereditarian opinions of many biologists and the environmental approach of most physicians would increase during the next decade and further inhibit communication and cooperation between these two professional communities.

PHYSICIANS AND EUGENICS

The general concern about the increasing numbers of socially unfit individuals in the United States continued during this decade. The nature-versus-nurture debate raged in this arena as social scientists presented evidence that the unfortunate lower classes were limited primarily by their environment, and thus social progress required better housing, food, and education. Biologists and many physicians argued instead for a strict hereditarian model of personality development. Expenditures of public funds to improve the lot of poor people would make no difference in subsequent generations because these traits were fixed by heredity and would be unchanged by improved living conditions.*

From the first year of the new century, physicians expressed increasing anxiety about the uninhibited reproduction of socially backward people. One described human propagation as "a mere bungle left to chance" (Surber, 1900). Another wondered why there was such "utter disregard of all laws of development and improvement among men" (Crowell, 1900). If personality traits were controlled by heredity, then being wellborn was of utmost importance for personal happiness and success, and for the improvement of society in general.

The question of whether physical disease was governed by heredity was quite controversial at this time, as noted previously. The bulk of dysfunction was felt by many physicians to be the result of environmental factors. But in a few specific instances heredity did appear to play an important causal role. Huntington chorea was repeatedly observed in several generations of parents and their children. As its course seemed unrelated to external agents, King concluded that it had a significant hereditary component that produced disease. He therefore advised against marriage by individuals from families in which chorea was present (King, 1906). When congenital cataracts were recorded in forty-three individuals in several generations of another family, it was declared to be "a social crime to bear more children in such marriages" (Dean, 1903).

But the major concern of physicians in this regard was the children born with seriously impaired mental or social abilities. There were alarming increases in the numbers of "insane, criminals, idiots, inebriates and tramp-paupers" (Johnson, 1905). H. L. Hunt observed that "there are probably no human characteristics so baneful and readily transmissible as idiocy, imbecility and insanity." He felt that it was a disgrace to modern civilization that this continued to occur (Hunt, 1910). Numerous family histories were presented in the medical literature as evidence for the hereditary nature of mental illness and retardation (Hymanson, 1902; Murdoch, 1902–3).

The proposed treatment plans for persons deemed socially unfit involved extensive regulation of human reproduction, in these cases by the "good Mother state" (Johnson, 1909). Public education on the potential for producing impaired children could be offered to prospective parents (Crowell, 1900; Morrow, 1907). Legal restrictions on marriages of individuals with socially undesirable characteristics could be

*For general review see Ludmerer (1972) and Kevles (1985).

enacted, but they might result in an increased number of illegitimate offspring. Contraception was also suggested when the mother had a chronic disease (Johnson, 1905). Isolated colonies for affected persons were advocated repeatedly as humane means to provide public support for individuals who could not function independently and at the same time segregate the sexes to prevent transmission of unacceptable traits. If all else failed, enforced sterilization was recommended to break the trail of repetitive transmission of these "hereditary" social disorders (Lambert, 1899–1900; Crowell, 1900; Hymanson, 1902; Southworth, 1908; Ounton, 1909; Johnson, 1909).

State legislatures began to consider laws for the eugenic sterilization of unfit individuals during the 1890s, although none were actually enacted into law until 1907 in Indiana. Physicians at state institutions had begun their own sterilization programs long before. F. H. Pilcher at the Kansas State Institution for Feebleminded Children castrated forty boys and fourteen girls before public outcry forced him to cease (Ludmerer, 1972). In a similar fashion, I. N. Kerlin had organized a castration program at the Pennsylvania Training School for Feebleminded Children in 1889. By 1899, H. C. Sharp at the Indiana State Reformatory had perfected the technique for vasectomy of male inmates. It involved cutting and tying a small segment of the vas deferens. The procedure was easy to perform and relatively painless. Dr. Sharp performed the procedure on a total of 465 male prisoners by 1907 (Kevles, 1985).

Release from state institutions was sometimes contingent on sterilization. Wisconsin had no legislation to compel anyone to accept this operation, but W. F. Becker reported that he had offered release to inmates of facilities for retarded persons if they or their families agreed to the operation (Wilmanth, 1910–11).

As experts on human disease, physicians were identified as the professionals who could best advise individuals and society on the means to prevent these social evils, whose increase threatened the very future of U.S. civilization. When physicians became more familiar with the mechanisms of heredity, they could then "enlighten and hygienically advise people as to the conditions which fit or unfit man and women for parentage" (Morrow, 1907). Graves predicted that physicians would soon decide who was genetically fit, and he believed that therefore "it becomes incumbent on the physician to familiarize himself with the important and rapidly progressing science of heredity" (Graves, 1910). Physicians were enlisted to become "conservors of the race" (White, 1909).

One of the earliest professional organizations of scientists and physicians which was involved in the issue of genetics and society was the American Breeders Association (ABA) (Kimmelman, 1983). By 1906 many leading biologists had become convinced that numerous socially important traits were influenced by heredity. The possibility of improving the human race through selective breeding appeared desirable, urgent, and technically feasible (Ludmerer, 1972; Ravin, 1985). D. S. Jordan, an ichthyologist, headed a Eugenics Committee formed within the ABA during its 1906 meeting. His initial discussion of the proposed role for the committee emphasized the social importance of heredity. Because social progress required better people as well as better living conditions, he proposed, society must do its utmost to promote the health of people with "good inheritable qualities." For those with "innate inferiority," the hope for society was that they should have no opportunity to transmit these qualities to their offspring (Jordan, 1907).

The Eugenics Committee members agreed that their objectives would be to: (1) investigate and report on heredity of human race; (2) devise methods to record the value of blood of different groups of people; (3) emphasize the value of superior blood and the menace to society of inferior blood; and (4) suggest methods of improving the heredity of family, people and race (Jordan, 1909).

The members of the committee also believed that knowledge of heredity when applied to human society could prove beneficial. Ward urged caution on the premature public discussion of the facts of human heredity before they had been carefully substantiated (Ward, 1907). Alexander Graham Bell agreed that better information was required, but that then public education on heredity should be encouraged by the professional community to promote voluntary compliance with recommendations on the factors that contribute to healthy offspring. He argued that racial improvement should be "persuasive rather than mandatory" (Bell, 1907). The major problem for society, according to Bell, was "to increase the quantity and quality of the desirable element and raise the general average of desirableness in the whole community." He urged therefore that the selective reproduction of superior couples would do more to better society than misguided attempts at preventing union of the less fit (Bell, 1909). Henderson was not convinced that this course was feasible. He worried that fit couples might be asked to make "unreasonable sacrifices in a hopeless competition with the unrestrained appetite of the unfit and undesirable." He believed that social isolation or sterilization of persons who were mentally ill or retarded or epileptic was a more realistic approach to social improvement (Henderson, 1909).

The sole physician on the committee, C. E. Woodruff, believed that the selection of marriage partners was governed by laws that were poorly understood by science. In general humans tended to mate with their opposites, thus producing offspring near the average for the population. He proposed that such matches be allowed to occur. The characteristics of the race would then be adjusted over time by the survival of the fittest. From his perspective, social misfortune was primarily the result of environment, not heredity. In his opinion, impairments were usually the result of bad food, clothing, or housing. His definition of eugenics involved the rearing of healthier children, not the selection of marriage partners (Woodruff, 1907).

But the hereditarian opinion of the majority of committee members prevailed. By 1910 C. B. Davenport had become the most vocal member of the Eugenics Committee. He was convinced that the Mendelian laws applied to humans as well as animals. Through selection, one could therefore produce "this or that character at will" (Davenport, 1907). Impairments were "inevitably transmitted in the germplasm and were apparently being reproduced faster than the more normal characteristics." In his opinion, many, "if not all, of the physical, intellectual and moral characteristics of man" were inherited (Davenport, 1909). He was certain that improving the human race was both urgent and technically possible (Ludmerer, 1972), and that there could be no more important goal than the enactment of genetic truths into law (Rosenberg, 1961). He proposed that good protoplasm was our "most valuable natural resource." Hence the most urgent public duty was "to maintain it and transmit it improved to subsequent generations" (Davenport, 1909).

Although Davenport, along with other biologists within the genetics community, urged physicians to become more involved in this area of human research and social betterment, he broadcast conflicting messages as to the ultimate goal of this project. (This campaign, sponsored by the Eugenics Record Office after 1910, is considered in more detail in the following chapters.) In one article in *American Breeders Magazine*, for instance, he urged careful and cautious study: "When precise laws of heredity shall have been determined for many human characteristics, it will be time enough to instruct people in regard to them. Premature attempts at instruction will bring the whole business into deserved reproach . . . Our greatest danger is for some impetuous temperament who, planting a banner of Eugenics, rallies a volunteer army of Utopians, free lovers and muddy thinkers to start a holy war for the new religion" (Davenport, 1910a). Later in the same year in the same publication, however, he urged public education on eugenics to arouse the populace and to promote legislation to restrain matings of "idiots, low-imbeciles, incurable and dangerous criminals—by isolation or sterilization." He encouraged society to protect itself by eliminating the "hideous serpent of hopelessly vicious protoplasm" (Davenport, 1910d).

These public opinions expressed by the leaders in the genetics community set the tone for thinking on the importance of human heredity in many areas of society. Davenport himself would become the key figure in U.S. human genetics after 1910, as other investigators focused on research involving heredity in plants and animals. His opinions powerfully influenced the medical community as well. The Eugenics Record Office (ERO) rapidly became the leading human genetics research institution in the United States. The publications and lectures by ERO workers defined the state of the art for the application of heredity to medicine and society at large. The ERO was the voice of authority.

CONCLUSION

By 1910 a unified theory of heredity had been postulated based on experimental studies in plants and animals. The new genetics successfully combined Mendel's laws of inheritance with the segregation of chromosomes and presumably the genes carried by them. Although not universally accepted within the biologic community, it had become the predominant theory in the United States to explain many observations on heredity.

Attempts to apply this model to human beings had just begun. There was no general consensus that such mechanisms even existed in humans. Some authorities in the biologic sciences believed that few important human characteristics were inherited, while others were convinced that virtually all human traits were controlled by unit characters that could be transmitted from one generation to the next.

Within the U.S. medical community there also existed a wide range of opinion on the importance of the new genetics. Many physicians were unaware of the new discoveries and discussed heredity in terms current in 1850 (Noble, 1909). Others were enthusiastic about the new information and argued that "no physician can afford to be ignorant of the matter, and it must be taught to medical students" (Mendel's law, 1907). It seemed, nevertheless, a most difficult and confusing area for study. J. B. Huber summarized the opinion of many physicians of his day:

> It is now generally agreed that this law [of Mendel] is up to the present time the most correct exposition of the phenomenon of heredity that we have . . . Ten years hence, we will no doubt be as familiar with it as we are now with the most obvious of nature's laws; so that we had best, as physicians, grasp it as soon as we can for the reason that it will enable us to understand the better many puzzling cases in our practice. But it is a very difficult matter to comprehend. I have for my part been pegging away at it for some time; and am coming just to see daylight regarding it. (Huber, 1908)

FOUR

GENETICS IN THE SPECIALTIES
Dermatology, Opthalmology, and Neurology

The maturation of theories on genetics affected the practices of different physicians to varying degrees. Three medical specialties—dermatology, ophthalmology, and neurology—developed along parallel lines in the United States after the Civil War. They viewed themselves as scientific disciplines because they followed the European model, which attempted to integrate laboratory study with clinical observations to accurately diagnose and treat specific disease conditions.

The process of professionalization occurred within the three specialty areas in roughly analogous fashion at about the same time during the latter half of the nineteenth century. A staged evolution permitted recognition of the specialty as a defined area of investigation and medical practice which was performed by carefully trained clinicians who were recognized as such by their professional colleagues. In the earliest stage of each specialty's development, a few individual general practitioners became particularly interested in diseases of the skin, eye, or nervous system. Their example may have encouraged younger medical graduates to travel to European centers for one or two years to study with the world leaders in their specialty of interest. There the U.S. students attended lectures, worked in clinics to observe patients, and performed some laboratory research, such as pathologic study of tissue from patients with particular diseases. The Americans then returned to the United States, designated themselves as specialists, and attempted to establish clinical practices dealing as much as possible with patients having problems in their specialty area. Recognition and legitimization of their status as specialists occurred when they were able to gather support from the community to establish specialized clinics or hospitals for the care of their patients. These institutions then served as centers to teach other practitioners the skills necessary for dermatologist, ophthalmologist, or neurologist.

Specialty textbooks also disseminated new knowledge to the general medical community, and medical periodicals publicized important new findings within each specialty field. The establishment of faculty positions at medical colleges recognized the outstanding specialists as leaders in the field, dedicating themselves to further investigation and teaching of new students. Finally, each specialty eventually formed its own professional organization, which held annual meetings to encourage the communication of new ideas. Because membership was strictly regulated by the founding specialists, an invitation to join implied that the applicant was considered competent by the recognized national leaders in a field. The professional model for the specialist as clinician, researcher, and teacher was an early example of the medical scientist who would eventually become the ideal physician during this modernization of U.S. medicine.

Despite their similar patterns of professional development, however, specialists in these three fields perceived the importance of genetics in daily practice quite differently and attempted to resolve the nature-versus-nurture controversy on the cause of human disease in distinctive ways.

DERMATOLOGY

The modern era of dermatology began in Vienna soon after 1850, as Hebra and his colleagues developed new cytologic techniques that permitted detailed microscopic study of both normal and diseased dermal tissue. They attempted to correlate clinical skin lesions with alterations in the normal microanatomy of the skin, emphasizing the importance of local irritants as the primary cause of many skin disorders.

Several U.S. physicians had demonstrated a particular interest in skin diseases even before this time. In 1837 H. D. Bulkley and J. Watson established the Broom Street Infirmary for Skin Diseases in New York and lectured on dermatology at the College of Physicians and Surgeons. Another early dermatologist was N. Worcester, who taught at the medical school in Cleveland during the 1840s.

The specialty of dermatology did not organize in the United States until European-trained physicians returned home with the new ideas in the field and established specialized practices. In 1865 F. D. Weisse was appointed to lecture on dermatology at the University of New York City and F. Smith began teaching at Bellevue Hospital Medical College. In 1871 J. C. White was named professor of dermatology at Harvard and established the clinical service at Massachusetts General Hospital. Dermatology began in Philadelphia at the same time, when C. A.

Duhring was appointed professor at the University of Pennsylvania in 1875. The establishment of dermatologic medical journals and societies began in 1870 with publication of the *American Journal of Syphilology and Dermatology*, succeeded in 1882 by the *Journal of Cutaneous and Venereal Diseases*. The first formal professional organization was the New York Dermatological Society, founded in 1869, followed in 1876 by the twenty-nine-member American Dermatological Association, with White as president (Pusey, 1933; Parish, 1976).

In the U.S. medical literature after 1880, many reports on familial skin disorders suggested that heredity could play a role in causing dermatologic disease. In contrast, a general review of skin diseases by Duhring and Stelwagon (1886) mentioned heredity as only a minor factor in cases of psoriasis and ichthyosis. In other cases, among them a series of cutaneous disorders observed in successive generations of different families, a more direct type of hereditary influence was recognized (table 4.1). The existence of direct parent-to-child transmission was viewed by dermatologists before 1900 as convincing evidence that "heredity no doubt plays a very important part in disease production in such cases" (Gilchrist, 1897).

In other types of skin disease, a disorder in one family appeared to demonstrate a different pattern of heredity than the same disorder in a second. Epidermolysis bullosa frequently involved only one child in large families with unaffected siblings, parents, and other relatives (Elliot, 1900; Hilton, 1905; Engman and Mook, 1906, 1910). The dystrophic form of the same disorder, however, was observed in five successive generations, with direct transmission of the character from parents to children (Engman and Mook, 1906). The interpretation of such familial clusters of dermatologic disorders changed dramatically

Table 4.1. Familial Dermatologic Disorders Reported, 1883–1913

Disease	Reference
Ectodermal dysplasia	Guilford, 1883
Tylosis palmaris et plantaris	Ballantyne and Elder, 1896
	Bromwell, 1902–3a
Dystrophia unguium et pilarum	White, 1896
Parakeratosis	Gilchrist, 1897
Prurigo of Hebra	Whitehouse, 1906
	Bamberg, 1913
Keratosis palmaris	William, 1907
Psoriasis	Engman, 1913

over the next decade: when three generations of another family were reported to have epidermolysis bullosa, the underlying cause was attributed to congenital syphilis, passed from parent to child through successive generations (Ravogli, 1917).

Xeroderma pigmentosum was another familial disorder that often affected several children in one generation (Hyde, 1903). Early reports in several instances noted that the parents in such families were often related (Taylor, 1888; Hutchins, 1893), suggesting that consanguinity might play some role in producing symptoms of the disease. Xeroderma was therefore called a "family disease" (Broyton, 1892–93), but was it hereditary?

The transmission of ichthyosis was even more complicated. The trait could be transmitted directly from parent to child, or it could skip a generation to reappear in a grandchild. It could also appear in a collateral branch of the family among the cousins (Mussey, 1895). Three successive generations of affected males were noted in one family (Shoemaker, 1907), while another family appeared to segregate the trait to males through unaffected females, the same pattern of heredity observed in families with hemophilia (Bromwell, 1902–3b). Affected females did occur, however, and they could transmit the disease to their sons, evidence for "the hereditary character of the disease" (Schwartz, 1909). Conversely, an isolated case with a negative family history was confusing. Wymer reported such an example and noted that "heredity may play a part in this condition in some cases, still I do not believe that it played any part in the case I shall report" (Wymer, 1908–9). By 1911 others were convinced that ichthyosis was not hereditary, but the result of a congenital infection transmitted from mother to child during the course of the pregnancy which eventually damaged the developing fetal skin (Young, 1910–11).

By 1920 histopathology had achieved great success in classifying many of the mechanisms that produced diseases in the skin. The number of infectious causes for skin diseases appeared great, as many diverse dermatoses had been shown to result from infection by bacteria or fungi. The logical union of etiology and pathology that now came to fruition seemed to produce an adequate explanation for skin diseases in general (Pusey, 1933).

There was no appreciation in the dermatologic literature of the developments within the science of genetics after 1900. The direct transmission of diatheses for skin disorders from parent to child, accepted as some evidence for heredity as a causative factor in such cases, was not supplanted by the new genetics. Instead, external irritants and infections appeared to provide mechanisms for pathology which fit the observed clinical facts.

OPHTHALMOLOGY

Ophthalmology was one of the earliest medical specialties to develop in the nineteenth century. By 1820 in Austria and Germany specialized eye hospitals had been established. Prominent teachers such as Beer, von Walther, and von Graefe were appointed to professorial positions in the medical schools, and the *Journal für Chirurgie und Augenheilkunde* had been launched to communicate new developments within the field of ocular disease (Shastid, 1917).

The first U.S. physician to specialize in eye diseases was Elisha North of Connecticut, who established his own eye hospital in 1817, a model for patient care and investigation of eye diseases. The first fully trained U.S. ophthalmologist was George Frick of Baltimore. After study with Beer in Vienna, he returned home to establish his practice in 1819, developed a large patient population, and reported his experience in *A Treatise on Diseases of the Eye* published in 1823.

Other young physicians followed a similar path of European study before establishing ophthalmologic practices in larger U.S. cities, among them Delafield and Rodgers, who founded the New York Eye and Ear Infirmary in 1820, and H. W. Williams, who initiated a lecture series on eye diseases at Harvard in 1850 and by 1869 was appointed to the newly established chair of ophthalmology at the medical school. But the first U.S. professor of ophthalmology was Elkanah Williams, in 1860 named professor at Miami Medical College in Cincinnati. Other early teachers of ophthalmology included E. L. Holmes at Rush Medical College in Chicago and W. F. Norris at the University of Pennsylvania (Shastid, 1917; Albert and Scheie, 1965).

The first publication in this new specialty was the *American Journal of Ophthalmology*, which appeared in 1863 but survived only three years. The bilingual German-English *Archives of Ophthalmology and Otology* began publication in 1869, with editorial offices in New York and Berlin, but by 1879 the two specialties had grown to such an extent that separate *Archives of Ophthalmolgy* and *Archives of Otology* were established. The *American Journal of Ophthalmology* was revived in 1884, and two newer journals appeared as the field expanded: *The Ophthalmic Record*, in 1891 and *Ophthalmology* in 1904.

The American Ophthalmological Society, founded in 1864 with

H. W. Williams as its first president, served as a model for the organization of other specialty associations over the next several decades. It functioned primarily as an important forum for the communication of new developments among practitioners of the specialty (Shastid, 1917; Albert and Scheie, 1965). Another significant opportunity for professional interaction was the series of congresses on eye diseases held in Europe and North America. The First International Ophthalmological Congress in Brussels in 1857 drew about 150 delegates, but apparently none from the United States. At the Fourth Congress in London in 1872, Williams described his experience with corneal sutures, the first major presentation by a U.S. ophthalmologist in this setting. The status of U.S. specialists had improved to such an extent that the next congress was held in New York in 1876, but it was primarily a national event. Of 106 delegates, only nine came from Europe. Subsequent meetings focused on the bacteriology of many eye diseases. Also described were new cytopathologic techniques that permitted detailed examination of disease processes within the different structures of the eye (Duke-Elder, 1958).

Although a general review of eye diseases at this time made no mention of heredity as a factor in causing ocular pathology (Norris, 1886), by 1911 Howe was able to claim that "in the whole range of medicine, probably no better examples of heredity can be found than those which show themselves in the eye" (Howe, 1911). From the earliest years of the nineteenth century, practitioners had reported the familial occurrence of eye diseases, which suggested that heredity did play a role in causing them. Kemp described a family with hereditary blindness in one of the earliest U.S. medical journals, the *Baltimore Medical and Physical Recorder*. At least three generations were affected. Nine of eleven children in one generation had the disorder and then passed it to some, but not all, of their offspring (Kemp, 1809).

The first use of a pedigree or family tree to illustrate heredity was presented in a report on color blindness. Five generations appeared to be afflicted, and about 90 percent of the affected were males. Females transmitted the character, although they were usually unaffected themselves. The pedigree was said to demonstrate "the overleaping of one generation by this hereditary peculiarity of vision" (Earle, 1845). Later reviews noted many families with similar male-predominant features. Affected females did occasionally appear, especially when individuals from both maternal and paternal lines were affected (Jeffries, 1879). One large family had color blindness in both lines and produced at least seven affected females in two successive generations (Reber, 1895).

The frequency of the trait varied among different populations. About 3 percent of white American males were affected and only 0.006 percent of the females. Other studies found 1.6 percent of black males affected (Burnett, 1879), and 1.8 percent of Native American males (Fox, 1882). The management of this disorder was not, however, believed to be hopeless, even though it was a hereditary problem. Burnett suggested that "the fact that females are afflicted less than males is most probably due to the circumstance that their faculty for color is more highly developed, and has been transmitted as a sexual proclivity from mother to daughter. It seems therefore quite reasonable to suppose that if boys could have their color-sense educated to the same extent as girls, and the process were continued through a number of generations, the defect of color blindness would in the course of time disappear" (Burnett, 1882).

The heredity of optic nerve atrophy was somewhat similar. In two large families direct parent-to-child transmission was observed in three successive generations. In other branches of the same families, the trait appeared to be transmitted through two unaffected women, who eventually produced many affected children of both sexes (Norris, 1880–84). In another family, five successive affected generations were observed, as well as one unaffected woman with two affected children (Norris, 1882). A six-generation family was studied by Gould, who noted that after the second generation, the disorder was transmitted only through unaffected females (Gould, 1893a, b, c). The same pattern of unaffected females and affected males was noted in other families of four or five generations (Posey, 1898; Hansell, 1900; Mix, 1903; Bruner, 1912), the identical form of inheritance observed in hemophilia and color blindness.

Glaucoma appeared at a young age among individuals in certain families. Three successive generations were affected in one large family (Howe, 1887). Pusey observed that the repetitive appearance of such a character indicated "that heredity was at work in causation of the disorder" (Pusey, 1892). Another family demonstrated segregation from parent to child through five generations (Harlan, 1898). By 1914, however, the report of a family with direct heredity failed to elicit any comments on mechanisms that might explain the transmission of such a disorder from one generation to another (Calhoun, 1914).

Iridemia and coloboma irides, also reported to be familial disorders, were observed in three successive generations of one family (De-Beck, 1894). In another case, direct transmission through three generations was noted again, but in one branch of the family two unaffected first cousins married and produced four unaffected children, one of whom then had a son affected with the disorder (DeBeck, 1886). Van der Hoeve, reporting an affected mother and two affected children, suggested that this could be a hereditary disturbance of normal embryonic development of the eye (Van der Hoeve, 1909).

Many other eye disorders also demonstrated the pattern of direct heredity from parent to child. Microcornea was observed in three successive generations of two families (DeBeck, 1900). Cataract was a common disorder that appeared shortly after birth among individuals in many families reported at this time, and congenital cataract was observed in three or four successive generations of several families (Williams, 1880; Wilson, 1891; Hansell, 1895–96; Millikin, 1904). In one series that studied the trait in a large number of families, ten of twelve children in a single generation were known to have cataracts. Clearly this disorder was familial, but "what is the actual part played by heredity?" (Wood, 1906). Other large affected families were reported after 1915, but there were often no comments on the possible causes of such a disorder (Parker, 1916a, b).

Direct inheritance was not always evident in other familial cases of cataract. A brother and sister who were afflicted had unafflicted parents who were first cousins (Alt, 1887). In another example, the parents were unaffected and unrelated, but three of six children were affected (Wood, 1906). And in a third family with unaffected parents, five of eight children were affected (Cheatham, 1903). The absence of direct parent-to-child transmission in these cases argued against heredity as a causative factor in producing this disorder. As Powers observed: "I have recently seen in my service the rare coincidence without heredity of three cases of congenital cataract in both of the eyes of the children in one family" (Powers, 1892). Another family demonstrated the transmission of the character from a father to three children, and the occurrence of the same disorder in two cousins as well (Dickey, 1898). By 1909 a different interpretation of these observations was made when a family with unaffected parents and three affected children was reported. "The occurrence of three cases in a family would seem to lend some weight to the well-established idea that the influence of heredity is a potent one in its etiology" (Robin, 1909–10).

A similar pattern of inheritance was observed in many families with ectopia lentis. Direct transmission of the character through two or three generations was quite common (Keyser, 1874; Bryant, 1892; Parker, 1898). Hence, heredity appeared to be "the most prominent etiological element in the production of the condition" (Bryant, 1892). A large number of children within the same generation were sometimes affected: in one family seven of nine children (Tiffany, 1895b), in another, six of eight (Miles, 1896). Lewis studied the trait in a large family of six generations and was able to examine patients from four successive generations to document direct parent-to-child transmission (Lewis, 1904). But anomalous cases did occur. In one family cousins were affected while everyone else was not (Thompson, 1887). In another example the parents were unaffected, but several of their children were afflicted (Tiffany, 1895b). Parents in a third family who were first cousins were unaffected, but three of their seven children were affected (Cheatham, 1903). The physicians reporting these cases made no attempt to provide a hereditary explanation for them.

Other common ocular disorders also recurred in successive generations of certain families: refractive error was evident in three generations (Frank, 1903–4), and astigmatism was also evident in five other families (Howe, 1911).

The inheritance of retinitis pigmentosa (a degenerative disorder resulting in blindness) was very confusing. In one series of nineteen families about 20 percent of the parents were related, suggesting that consanguinity might be important in producing symptoms of the disease. There were instances of direct parent-to-child transmission for two, three, or four generations. But in other cases, the parents were normal and yet had affected offspring (Ayres, 1886). Later studies of other interrelated families demonstrated the same pattern and suggested the importance of consanguinity as a factor in this condition (Coleman, 1889; Belt, 1896; Sweeney, 1912). Even without evidence of direct transmission, DeBeck was willing to state that "some sort of hereditary influence is responsible for this condition" (DeBeck, 1897).

Between 1900 and 1915 several extensive review articles examined the role of heredity in causing diseases of the eye. Weeks believed that all parts of the eye could be afflicted by heredity (Weeks, 1903). Three articles at this time summarized the eye diseases that appeared to be caused by heredity (table 4.2). The authors were unable to explain hereditary mechanisms, but there was general agreement that consanguinity seemed to "accentuate the hereditary tendency" (Weeks, 1903). Specific eye disorders did often appear in the offspring of parents who were related by blood. Dean noted two first-cousin marriages within one family in which four of seven children were affected by retinitis

Table 4.2.   Hereditary Eye Diseases Reported before 1910

Albinism	Glaucoma
Amaurotic family idiocy	Megalophthalmia
Aniridia	Microphthalmia
Aphakia	Myopia
Astigmatism	Night blindness
Blindness	Nystagmus
Cataract	Optic nerve atrophy
Coloboma	Ptosis
Color blindness	Retinitis pigmentosa
Corneal degeneracy	Word blindness
Ectopia lentis	

Source: Data from Weeks, 1903; Libby, 1909; Franklin, 1913.

pigmentosa. Albinism was another disorder frequently observed in children born of healthy but related parents. Glaucoma and cataract followed similar patterns. In one large intermarrying family, forty-three people with cataracts were reported (Dean, 1903). Consanguinity was also observed in families with many other eye diseases. First-cousin parents had three children afflicted with macular degeneration (Fein-gold, 1916). Libby noted that such a pattern implied that a genetic factor must be involved in causing the disorders. He proposed that Mendel's theory of the intensification of qualities through inbreeding might be applicable in such cases of human disease (Libby, 1908). But other genetic theories were also considered relevant. Loeb reported 496 cases of familial eye disease which he classified as forms of direct, indirect, or collateral inheritance as defined by Ribot's 1875 theory (Loeb, 1909).

The usefulness of Mendel's theories for understanding human inheritance was appreciated by more eye physicians after 1910. Fischer not only believed that Mendel's laws held true for the transmission of human physical defects but also proposed that it was possible to predict with near certainty the results of matings between two parents with specific eye diseases (Fischer, 1914). Danforth agreed and encouraged ophthalmologists to report cases of hereditary eye disease, including both affected and unaffected family members (Danforth, 1916). Howe observed that because eye physicians regularly recorded the family history in cases of eye disease, the advent of Mendelian inheritance would make it possible to classify these eye diseases based on their pattern of heredity. He believed that the appearance of certain disorders in only one sex could be explained on the basis of the segregation of the sex-determining chromosome (Howe, 1917).

Table 4.3.   Eye Disease and Mendelian Inheritance before 1920

Disease	Reference
Dominant	
Cataract	Church, 1914
	Danforth, 1914
	Ziegler, 1915
	Buxton, 1916
Coloboma	Church, 1914
Distichiasis	Church, 1914
Ectopia lentis	Church, 1914
	Buxton, 1916
Hemeralopia	Church, 1914
Retinitis pigmentosa	Church, 1914
Strabismus	Church, 1914
Ptosis	Briggs, 1918
Recessive	
Albinism	Church, 1914
Cataract	Jones and Mason, 1916
Sex-linked	
Color blindness	Church, 1914
Nystagmus	Church, 1914

Church was among the first ophthalmologists to publish a classification of eye diseases in terms of their Mendelian inheritance patterns. Dominant traits such as ectopia lentis and cataract were fairly common. Recessive disorders were less frequent but were seen more often in consanguineous marriages. Sex-linked characters did indeed follow the segregation of the sex-determining chromosome (Church, 1914). By 1920, the hereditary classification of a number of eye diseases was famliiar to U.S. ophthalmologists (table 4.3).

Strabismus, often observed in successive generations, fit the pattern of a dominant trait (Church, 1914). Coloboma irides (congenital fissure of the iris) in father and son was interpreted by Davenport to show a hereditary factor in action. He urged those with coloboma not to marry because their children were all liable to be affected (Lewis, 1915). Church, who interpreted this familial pattern to be most consistent with the transmission of a Mendelian dominant trait, also believed that retinitis pigmentosa was for the most part a dominant character. Although he did note that recessive disorders were more common in families in which the parents were related by blood, he did not specifically analyze the anomalous indirect cases of retinitis noted previously

(Church, 1914). (More recent studies have shown that retinitis pigmentosa is a heterogeneous group of eye disorders. Examples of dominant, recessive, and sex-linked heredity are now well documented. Dryja, Hahn, and McGee, 1991; Humphries, Kenna, and Farrar, 1992). Hemeralopia (night blindness) also demonstrated a direct pattern of heredity in many cases. Males and females were equally afflicted. In one large family that demonstrated the trait through five generations, all affected individuals had an affected parent (Bordley, 1908), a pattern again consistent with Mendelian dominant inheritance (Church, 1914). Ptosis (a drooping upper eyelid), which appeared in a six-generation family with sixty-four affected people, also seemed to segregate as a dominant character (Briggs, 1918). Spiece believed that Mendel's laws might pertain to the segregation of ectopia lentis (dislocated lenses), but he was unable to apply them to specific families (Spiece, 1919). Church and Buxton agreed that this was most likely a dominant disorder (Church, 1914; Buxton, 1916).

The analysis of the segregation of cataracts reveals the uncertainty among physicians as to the relevance of the new genetics to clinical practice. When a family was reported with an affected father and four of eight affected children, raising the question of why only every other child was afflicted, the author concluded that to confront this issue would "open a line of thought and discussion on heredity and eugenics clearly out of place in this paper, although Mendel's laws and suggestions must command our attention and add to our wonder" (Campbell, 1913). Church, however, concluded that this pattern of segregation indicated that cataract was a dominant trait (Church, 1914), as did Danforth, who studied a family with three successive affected generations (Danforth, 1914), and Ziegler and Buxton, who reported large affected families (Ziegler, 1915; Buxton, 1916). A more formal reevaluation of the heredity of human cataract by Jones and Mason of the Bussey Institute revealed a high incidence of consanguinity among the parents of affected children in such families. They concluded that certain types of the disorder most likely segregated as a Mendelian recessive trait rather than as a dominant one (Jones and Mason, 1916).

Nystagmus, a rapid, involuntary oscillation of the eyeballs, was transmitted in quite a different fashion. An interesting family was reported in which an affected man, married twice to unaffected wives, had fathered unaffected daughters, who then produced affected sons. One example of direct father-to-daughter transmission of the character (Frank, 1903) was believed to represent a Mendelian sex-linked pattern of inheritance (Church, 1914). The genetic mechanism for the transmission of color blindness through the mother to affected sons was understood to be similar: a double homozygous dose would be required to produce symptoms in the female, while only a single heterozygous dose would be required to produce an affected male (Church, 1914).

Opthalmologists debated the relative importance of hereditary constitution and environmental factors in producing disease, as did all other specialists, each in their own arena. At a meeting of the Ophthalmological Section of the American Medical Association, Howe stated that "we ophthalmologists have been content thus far with reporting family histories without attempting to relate those histories to other facts now well established by geneticists." In his opinion, the knowledge of whether a particular character was dominant or recessive was important, as the physician counseled men and women on the advisability of marriage in certain families where such eye diseases were present (Howe, 1918).

The definition of the role played by heredity was twofold, however. A predisposition could be transmitted, which later became manifest as physical disease, or the disease itself could be transmitted, as in the case of congenital syphilis (Lanier, 1919–20). Howe summarized the perplexing nature of the cause of these eye diseases when he observed that it was truly difficult to know whether a particular eye defect was hereditary or the result of an acquired infection (Howe, 1919a). Not limited to ophthalmology, this quandary reflected the state of the understanding of pathologic mechanisms in all areas of clinical medicine of this era.

NEUROLOGY

While the professionalization of neurology as a clinical specialty paralleled developments in dermatology and ophthalmology, the field used heredity to explain processes of disease in quite a different manner. There was general agreement from the earliest days of the specialty that much disease in this realm of medicine was hereditary. Neurologic disorders were said to "obey the laws of inherited diseases" in that derangement in one part of the nervous system might be transmitted to the progeny in a different form. The factor inherited was felt to be a "family proclivity to nervous disorders, in one case idiocy, in another mania, in another convulsions" (Rogers, 1869). The spectrum of these hereditary predispositions ranged from catalepsy, ataxia, and epilepsy

Table 4.4.   Hereditary Neurologic Diseases Reported in 1902

Acute poliomyelitis	Hemiplegia
Amaurotic family idiocy	Huntington chorea
Amyotrophic lateral sclerosis	Hypochondria
Angioneurotic edema	Hysteria
Astasia abasia	Intracerebral aneurysm
Ataxia paraplegia	Landouzy-Dejerine muscular
Cerebellar ataxia	atrophy
Cerebral apoplexy	Little disease
Cerebral diplegia	Migraine headache
Cerebral tumors	Paralysis agitans
Charcot-Marie-Tooth peroneal	Pseudohypertrophic muscular
atrophy	paralysis
Cretinism	Spastic paraplegia
Epilepsy	Spina bifida
Exophthalmic goiter	Thomsen disease
Hemiatrophic facialis	Torticollis
	Tuberculous meningitis

Source: Data from Krauss, 1902.

to murder, suicide, and "imbecility" (Rogers, 1869; Layton, 1882) (table 4.4).

The clinical specialty of neurology developed in Paris after 1850 with the pioneering work of G. B. A. Duchenne. He studied many patients with neurologic disease in various hospitals and introduced electrical stimulation of nerves and muscles to understand their concerted action. His student Jean Martin Charcot, appointed to the chair of neurology at the Faculty of Medicine at the University of Paris in 1882, organized a clinical service at the Salpêtrière with a neuropathologic laboratory in which the pathologic changes in nervous tissue could be correlated with the clinical symptoms observed in life.

German neurology arose during the last quarter of the nineteenth century, as Rudolf Virchow, Alois Alzheimer, and Franz Nissl developed improved techniques for studying pathologic changes in peripheral nerves, the spinal cord, and the brain itself. Their clinical colleagues Wilhem Erb, Nikolaus Friedreich, and Herman Oppenheimer described pathologic lesions characteristic of many clinical neurologic diseases (McHenry, 1969).

Neurology in the United States developed as a clinical specialty during the Civil War. The surgeon general at that time was W. A. Hammond, one of the first U.S. specialists in this area. He organized the U.S. Army Hospital for Diseases of the Nervous System in Philadelphia and encouraged physicians to study the patients with neurologic disease and injury. In 1871 he reported his experience in the nation's first neurology text, *A Treatise on Diseases of the Nervous System.* Because of Hammond's efforts, Philadelphia became the first center for intensive study of neurologic disease in the United States. S. Weir Mitchell worked at the army hospital and eventually cultivated an extensive clinical practice specializing in diseases of the nervous system. Hammond and Mitchell, who had both studied in Paris, attempted to correlate clinical neurologic symptoms with changes within the structure of the nervous system. The University of Pennsylvania established a Department of Neurology in 1871 with H. C. Wood as the first lecturer. His colleague C. K. Mills subsequently founded the clinical service at the Philadelphia General Hospital in 1877 (McHenry, 1969; DeJong, 1982).

Other U.S. physicians started neurologic clinics in New York, Boston, and Chicago. After a period of European study E. C. Seguin lectured on diseases of the nervous system at the College of Physicians and Surgeons beginning in 1870. J. J. Putnam, also trained in Europe, established the neurology service at Massachusetts General Hospital in 1872 (McHenry, 1969). At the same time, J. S. Jewell was appointed professor of neurology at Northwestern University Medical School in Chicago (DeJong, 1982).

The profession organized itself as the American Neurological Association at a meeting in New York in 1875. Eighteen prominent neurologists had been invited to attend, and Jewell was elected the first president. The association met annually thereafter and provided an important forum for communicating new developments within the field, correlating disease symptoms and neuropathologic lesions (DeJong, 1965, 1982). The professional status of neurology was further enhanced by the initiation of the *Journal of Nervous and Mental Diseases* in 1874, which became the primary vehicle for promulgating new findings in U.S. neurology (Brandy, 1960; DeJong, 1965, 1982).

The physicians' understanding of hereditary mechanisms in neurologic disease paralleled developments within the new science of genetics. Before 1910 the traditional pattern of direct transmission of a character from parent to child was believed to be the best evidence for heredity as a cause of the disease in question. As neurologist H. A. Cottell commented, "Any nervous infirmity may be hereditary" (cited in Marvin, 1902–3). The application of Mendelian genetics in neurology, tentative at first, became more widespread after 1915.

The preeminent "hereditary" neurologic disease was unquestionably Huntington chorea. The earliest case reports by Waters called it "markedly hereditary." He was unaware of an affected person who did not also have an affected parent and grandparent (Waters, 1842). Subsequent families were observed with affected individuals in five successive generations, and physicians agreed that it was difficult to explain the presence of this disorder in so many generations except on a hereditary basis (Lyon, 1863). But transmission was not invariable. Some children of affected parents never developed any symptoms of the disease. In this fashion "the thread was broken," and subsequent generations were entirely free of the malady (Huntington, 1872; King, 1885, 1885–86). The symptoms of chorea usually did not appear until after age forty. Several families were reported in which apparently unaffected parents died before symptoms of the disease occurred but whose offspring developed typical symptoms later in their life, implying that the hereditary factor had indeed been transmitted (Sinkler, 1889; Tilney, 1908).

Collins believed that in such cases initially normal neurons were predestined to an early death. "The neuron is genetically lacking in the power which will enable it to exist as long as the ordinary ganglion cell" (Collins, 1898a). (The term *genetic* in this sense implies a developmental defect that eventually produces neurologic symptoms.) The importance of familial diathesis in producing chorea was emphasized in reports of numerous family histories in the medical literature before 1900 (Hamilton, 1908). King was certain that "so decided am I as to heredity being the great predominating, if not the only, factor in inducing the disease that I should consider a diagnosis of this form of chorea unjustifiable without a history of adult chorea in an ancestor" (King, 1906).

Of seventy descendants of another large family with direct heredity through at least four generations, thirty-three were afflicted with chorea (Hattie, 1909–10). Dana, who thought that the new ideas on heredity might be applicable in such cases, observed that "Huntington's chorea had always appealed to me as an excellent disease in which to work out and apply the Mendelian theory" (Huntington, 1910). Boyd agreed that this theory might be relevant (Boyd, 1913), and White eventually reviewed a large number of pedigrees and proposed that chorea acted as a dominant character (White, 1913). Further investigation of pedigrees from state mental hospitals by McKinnis (McKinnis, 1914) and from the Eugenics Record Office by Davenport and Muncey (1916) substantiated the opinion that Huntington chorea followed a dominant pattern of inheritance.

A similar pattern of direct inheritance was often observed in families with ataxia, the inability to smoothly coordinate voluntary motion. Four or five generations in succession were frequently affected, males and females about equally. But symptoms of disease were not always evident in those carrying the trait, so that unaffected people occasionally produced children with symptoms of ataxia (Brown, 1891–92; Neff, 1894–95, 1905).

Thomsen disease (or myotonia congenita, in which muscles develop spasms during voluntary actions) was another neurolgic disorder in which heredity was believed to be important (Mills, 1898). Occasionally an isolated case appeared in a family with no antecedent history (Jacoby, 1887), but more commonly four or five generations demonstrated direct parent-to-child passage of the character (Sedgwick, 1910; Steiner, 1915).

Neurofibromatosis (von Recklinghausen disease, a disorder in which benign tumors grow along peripheral nerves) also was evident in successive generations in particular families, an occurrence felt to "afford pretty conclusive evidence that heredity may play a part in the etiology of the affection" (Atkinson, 1875). While direct heredity appeared to be most common (Meek, 1905), examples of "skip generations" were noted in which individuals with no apparent clinical symptoms fostered children and grandchildren affected by neurofibroma (Harbitz, 1909). Preiser and Davenport examined twenty-seven pedigrees and found that the disease was usually directly inherited, affecting equal numbers of males and females. They concluded that the trait segregated in a dominant fashion (Preiser and Davenport, 1918).

Other types of neurologic disease that appeared to be transmitted directly from parent to child included the jumping Frenchmen of Maine syndrome, in which affected persons jumped uncontrollably when confronted with a loud noise. Because it recurred in two or three successive generations of certain families, it was believed to be "fully as hereditary as insanity, epilepsy or hay fever" (Beard, 1880–81). Apoplexy was another such trait (Harrington, 1885; Merzbach, 1911).

Before 1910 the presence of direct parent-to-child transmission was believed to represent the best evidence that heredity was an important factor in producing certain neurologic diseases. But such a theory did not explain, for example, three brothers affected with spastic paraplegia with six unaffected siblings and unaffected parents. This did not fit the expected pattern of "heredity, " and the etiology of this disorder was thus labeled "indeterminate" (Eshner, 1896–97). More frequently the same disease was observed in many families with up to five affected successive generations (Bayley, 1897; Spiller, 1902, 1910b, 1915b), apparently solid evidence for the "true hereditary character of this type of degeneration of the central nervous system" (Mason and Rienhoff, 1920).

Migraine (vascular headache) was another disorder that frequently recurred in families. Knapp described eight individuals affected in three generations (1907). Cases did occur without antecedents (Neu, 1911), but generally it was agreed that there was no "nervous disease which is transmitted from parent to child as often as migraine; no one in which direct heredity plays so important a role" (Andvist, 1913). The possibility that Mendelian inheritance could explain the familial nature of migraine was first considered by Buchanan, who reviewed the pedigrees of 127 families with this disorder and in an initial report concluded that migraine did segregate as a "Mendelian phenomenon" (Buchanan, 1920). Further consideration led him to reject this notion in a subsequent report in the following year (Buchanan, 1921), when he was no longer sure that Mendelian traits existed in human beings in the first place.

Familial tremor also appeared in successive generations of large families (Dana, 1887). White believed that such evidence of direct heredity indicated that it segregated as a dominant trait (White, 1913). On the other hand, periodic paralysis exhibited two distinct patterns of inheritance: direct heredity through three or four generations (Holtzapple, 1903–4), and heredity that induced a susceptibility to some toxic agent, which eventually resulted in symptoms of paralysis. This was the case in a family in which several unaffected females appeared to transmit the character to their offspring, and an affected female in the same pedigree also produced two affected sons (Atwood, 1912).

Defects in the embryologic development of the neural tube also appeared in several children of the same parents. In one instance, unaffected parents were related as first cousins and produced two children with spina bifida, a defect in the bones encircling the spinal cord (Holt, 1887). In another family three children were born with encephalocoele, the absence of the bony plates of part of the skull (Phillips, 1908). The absence of affected ancestors made it difficult, however, to determine whether heredity was a causative factor in such cases.

Pseudohypertrophic muscular paralysis (Duchenne disease) affected boys in successive generations of many families. One study of five pedigrees observed affected males in collateral branches; the character appeared to be transmitted through unaffected females (Poore,

1875). The lack of direct heredity raised serious questions as to the importance of inheritance in producing symptoms of this disease. "This peculiar affection is to a certain, probably we ought to say a slight, extent hereditary. It should more properly be called a diathetic disorder that may be peculiar to a family" (Bridge, 1885). In another example, Porter noted two affected boys in one family and reported that, "as there seems to be an hereditary tendency to the disease and as these cases occurred in the same family, I had expected to find something of this sort in the family history, but I have been unable to find a paralytic or neurotic history on either side of the family" (Porter, 1897). As to why only boys were affected, Russell thought that "it seems, indeed, extraordinary that sex should influence the onset of the disease" (Spiller, 1910c). The issue of heredity in this form of paralysis remained uncertain even after the application of modern theories of genetics. Camp concluded that "we are not dealing with a hereditary characteristic which could be transmitted as a unit character in the Mendelian sense." Instead, some "degenerative factor" was transmitted that somehow caused tissue destruction and clinical symptoms of muscle disease (Camp, 1916). And yet a similar disorder, facioscapulohumeral muscular dystrophy, appeared to involve both sexes and was transmitted directly from parent to child through four generations (Marvin, 1902–3).

Because the definition of heredity at this time called for the direct transmission of a character from parent to child, the familial nature of a disease did not necessarily mean that it was hereditary. A muscle degenerative disease, Charcot-Marie-Tooth peroneal atrophy, was reported in the males of a family through at least four and perhaps seven generations. Although unaffected females appeared to transmit the character in most lines of the family, one affected female did have an affected son (Church, 1906). In another family, unaffected parents had one affected son and two affected daughters, evidence to suggest that the disease was familial, not hereditary (Hatch, 1915). In contrast, the existence of some examples of direct parent-to-child transmission led Spiller to conclude that the disorder was both familial and hereditary (1915c).

The same type of controversy was evident with Friedreich ataxia, which often affected siblings. Parents were usually clinically well; hence, evidence of direct heredity was generally lacking. Occasionally the parents were consanguineous. All of these observations clearly indicated that this was a familial disorder (Smith, 1885; Griffith, 1888; Wells, 1888; Frazier, 1903–4) and that "previous reports of 'hereditary' ataxia are not strictly speaking, hereditary, for neither the parents nor any other ancestors were ataxic" (Sinkler, 1885). Or even more conclusively: "These cases tend to show that the term 'hereditary' ataxia is a misnomer and misleading" (Shattuck, 1888). In a large series of 100 families, only 6 instances of direct transmission were demonstrable. The question of heredity was therefore "left to our imagination." The occurrence of disease symptoms in children of the same family could equally be explained by the presence of environmental factors shared by all of them (Robins, 1906–7). It was noted repeatedly, however, that members of the family who did not suffer from ataxia often exhibited other symptoms of neurologic disease such as chorea, paralysis, hysteria, mental affections, or intemperance (Smith, 1888). This neuropathic tendency provided a "soil suitable for the development of ataxia" (Hunt, 1910), but how such a familial weakness was transmitted was inexplicable (Sinkler, 1906). Large families often had ataxic relatives in collateral branches, but again with unaffected parents (Fry, 1896; Brower, 1897; Kellog, 1898). This pattern of inheritance was finally interpreted by White as most consistent with the segregation of a Mendelian recessive character (White, 1913).

The same familial pattern was observed in cases of Oppenheim disease (dystonia musculorum deformans, a type of progressive muscle weakness); that is, affected individuals were born of unaffected parents, with no history of such disease in preceding generations. Coriat argued that this indicated that a "recessive factor was at work, according to the scheme of Mendelian inheritance" (Coriat, 1916).

Amaurotic family idiocy (Tay-Sachs disease) was another familial disorder recognized at this time. The first cases described before 1900 showed evidence of neuronal degeneration in the brain and spinal cord. Sachs called it an "agenetic condition, " implying a failure of normal embryonic development (1887). In all instances, parents of affected children were clinically healthy; several were consanguineous (Carter, 1894). Some suggested that the damage to the central nervous system might be the result of a toxicant, perhaps transmitted via the mother's milk (Hirsch, 1898; Hymanson, 1902). By 1900, other investigators had concluded that more than arrested healthy development was involved in producing clinical symptoms of this fatal disease. A destructive process appeared to be changing the neuronal architecture of the brain (Kuh, 1900). Perhaps an inherent biochemical property of the protoplasm of the cells resulted in the degeneration of the neurons (Davis, 1909). The mechanism controlling such a metabolic factor remained, however, unknown. McKee suggested in a prescient article that "it may be to the sperm cell or the ovum that we must look, and possibly to the chromosomes of these cells" (1905). The sexes were equally affected. Certainly there was a familial tendency in the disease, but was heredity an etiological factor? Frank believed that heredity was, in fact, of little importance. In a family with six children, only three might be affected, while the others escaped (Frank, 1906). The remarkable ethnic proclivity of the disease was also an enigma. Parents invariably were of East European extraction, most often Polish or Russian. "Why the biochemical property is confined to the protoplasm of the cells of little Hebrews no one has yet described" (Buchanan, 1907). The review by Sachs concluded that the disease was not due to toxicants in breast milk, because some affected children had been nursed by women outside their own family or had used only artificial formula (1915). Cerebrospinal fluid had been collected from affected children and injected into experimental animals. Pathologic examination of the brain and spinal cord revealed no apparent defect, again arguing against the existence of a neurotoxicant as the underlying cause of the disease (Welt-Kakels, 1917).

Sachs was convinced that heredity played an important role in this neurologic disorder, labeling it "the starting point for the future study of other hereditary and family diseases of the nervous system" (1910). Others noted that it had been "strangely neglected in the modern investigations on heredity. The fact that in some families only one case appears does not militate against the familial nature of the disease" (Coriat, 1913). The application of modern genetics to amaurotic family idiocy was described by Herrmann at a meeting of the American Pediatric Society, where he suggested that "inbreeding among Hebrew families could increase the likelihood that both parents would transmit the same character to some of their progeny." The mode of transmission for this character was best explained as a Mendelian recessive trait (Herrman, 1915a, 1915b). A few years later, Brandeis was able to explain the genetic mechanism in greater detail. Consanguinity would produce parents who were heterozygous for the trait but clinically healthy. About 25 percent of their offspring would be expected to be homozygous recessive and thus develop symptoms of the disease (Brandeis, 1918). Even this detailed explanation did not convince everyone in the field of the relevance of genetics for the disease. Epstein felt that its etiology was still "mysterious" and found it "difficult to understand why the sixth or seventh child only in a family is affected" (Epstein, 1917a), but he was willing to wait for further developments. In a subsequent review he suggested that in the future "the Mendelian

principle may perhaps supply the missing etiology" (Epstein, 1920).
Epilepsy was traditionally viewed as one of a series of neurologic disorders passed from one generation to another via a hereditary diathesis that could be excited by external factors to produce convulsive symptoms of the disease. At times, the familial nature of epilepsy was indeed remarkable. A woman with epilepsy bore nine children, all of whom developed seizures in childhood (Gray, 1879), and three successive generations of persons with epilepsy were observed in another large family (Sinkler, 1907). "The cause of epilepsy is presumably an hereditary taint, by which I mean particularly, that in tracing back in the ancestry of the patient we will almost invariably find that there has been an hereditary proclivity or predisposition to nervous disease upon slight exciting causes; that is, in other words, there is a family neurosis which has manifested itself in certain generations by insanity, idiocy, phthisis or inebriety" (Mann, 1883).

But the relative importance of direct heredity in causing epilepsy was controversial. Osler reviewed 126 pedigrees of affected people and found that thirty-six had some antecedent relative with evidence of "neurologic derangement, " such as insanity, epilepsy, paralysis, or hysteria. Only two people had mothers with epilepsy, and none had affected fathers. He concluded that direct inheritance was in fact rare, but that families with neurosis often produced children with epilepsy (Osler, 1892). A larger series of 1,300 pedigrees yielded similar results. Only 8 percent of the cases had one parent with epilepsy, and another 11 percent had a collateral affected relative; 8 percent had relatives with insanity, and 22 percent had relatives affected by alcoholism. In sum, one or more defective traits were observed in the families of 39 percent of all cases (Doran, 1903–4).

Subsequent surveys of large populations of individuals with epilepsy consistently found less than 10 percent of cases with an affected parent (Fairbanks, 1914; Thom, 1915; Burr, 1922). Hence it did appear that heredity (defined as direct transmission of the character from parent to child) played a rather minor role in causing symptoms of epilepsy (Kehoe, 1915).

Davenport and his colleagues at the Eugenics Record Office investigated 175 families with epilepsy and attempted to define the segregation of this character in Mendelian terms. They concluded that it acted like a recessive unit character (Davenport and Weeks, 1911). Similar results were obtained in a study of almost 400 families with family members at the New Jersey Village for Epileptics (Weeks, 1912, 1915). Many other neurologists agreed with this interpretation of epilepsy as

Table 4.5.   Neurologic Disease and Mendelian Inheritance before 1920

Disease	Reference
Dominant	
Huntington chorea	Shannon, 1913
	White, 1913
	McKinnis, 1914
	Davenport, 1915c
	Davenport and Muncey, 1916
Familial tremor	White, 1913
Neurofibromatosis	Davenport, 1918
	Preiser and Davenport, 1918
Recessive	
Epilepsy	Davenport and Weeks, 1911
	Weeks, 1912
	White, 1913
	Weeks, 1915
Amaurotic family idiocy	White, 1913
	Herrman, 1915a, c
	Brandeis, 1918
Friedreich ataxia	White, 1913
Oppenheim disease	Coriat, 1916

a recessive trait (Perry, 1911–12; White, 1913; Hubbard, 1914).

But not all investigators in the field were convinced. Clark argued that the data were inadequate to show that epilepsy segregated as a Mendelian unit character. He believed that epilepsy was most likely a symptom of different diseases that produced irritability of the central nervous system, rather than a single disease entity (1912). It had long been recognized that people with epilepsy were often born into families with histories of other neurologic or emotional symptoms: "That epileptic children should be born of neuropathic-degenerative stock, including alcoholics, is almost a foregone conclusion" (Clark, 1915). Epilepsy also appeared in families with traits such as drunkenness, mental illness, migraine, mental retardation, pauperism, vagrancy, or prostitution (Clark, 1915; Thom, 1915). Many physicians felt its appearance was evidence of a familial abnormality in the germ cells which resulted in the transmission of a predisposition to irritability of the nervous system. Environmental triggers then might produce symptoms of epilepsy when acting upon these susceptible individuals (Burr, 1922).

This type of information was used to support that argument that much familial neurologic disease was hereditary. A defect in the germplasm could be transmitted from parent to child to produce symptoms of disease with little, if any, influence from the environment (Davenport, 1912b). But surprisingly few specific neurologic disorders had a segregation pattern that was explained by physicians of this era in Mendelian terms (table 4.5).

Neuropathologic investigations defined specific lesions for many diseases of both the central and peripheral nervous systems, and attempts to improve the precision of diagnosis were ongoing. Knapp believed that the use of the term *hereditary* was often indiscriminate and confusing within the field of neurology. He observed that the data on the heredity of many nervous disorders were "absolutely unsatisfactory, if not utterly worthless . . . The whole teaching as to the etiological importance of heredity rests upon a mass of inaccurate, undigested data which often have no bearing upon the real etiology of the disease under consideration" (Knapp, 1907). The unsettled state of the importance of heredity in causing neurologic disease reflected the continuing controversy of nature versus nurture in the broader field of medicine at this time.

FIVE

THE ROLE OF GENETICS IN THE PRACTICE OF MEDICINE
1910 to 1922

MEDICAL EDUCATION AND GENETICS

The gradual evolution of U.S. medical education from vocational to scientific process paralleled changes in university science education in general after the Civil War. Leading undergraduate institutions implemented courses in physical and biological sciences as integral parts of the liberal arts curriculum (Brieger, 1983; Bruce, 1987). Graduate programs offered opportunity for specialized research in the sciences and encouraged students to continue their training at home rather than to travel to Europe for advanced experience (Bruce, 1987).

The incorporation of science into the medical curriculum was a much slower process, for the medical profession at large was not convinced that specialized training in biology, chemistry, and physics was important to produce a competent physician until the early 1900s.

Around this time an increasing number of physicians came to recognize that medical science was more pragmatic than they had assumed. Their enhanced ability to diagnose diseases accurately resulted from the application of discoveries made in the research laboratory, and they continued to hope that the skillful application of science would result in cures for their patients in clinical practice (Ludmerer, 1985; L. King, 1991). Because of this change in professional opinion, a basic scientific education came to be deemed essential for "everyone who would be considered a modern practitioner" (Ludmerer, 1985). This new emphasis in medical training had begun to encourage young doctors to think of themselves as medical scientists rather than as craftspeople.

The establishment of the Johns Hopkins School of Medicine in 1890 was an important influence in this gradual shift in thinking. From its earliest planning stages, Johns Hopkins was to be a medical institution with radically new traditions. Only the best students with college degrees were to be admitted. The mission of the school was to ensure that the medical profession comprised "learned men, not tradesmen" (Mall, 1987). An important goal of the new school was to combine medical research with clinical practice: basic research was encouraged within the clinical departments (Ben-David, 1960), and the students were encouraged to develop a spirit of inquiry that would carry over to their professional lives after graduation. This was to be an attempt to combine the German tradition of medical research with the English tradition of clinical teaching on the hospital wards (Pickering, 1976). The graded curriculum at Johns Hopkins—two years of preclinical work in the basic sciences followed by clinical rotations on all the specialty wards of the hospital—set the standard for medical education in the United States and resulted in a "veritable revolution" in medicine after 1890 (Shryock, 1953).

By 1900 two sorts of medical college had developed in the United States. Elite institutions followed the Hopkins model for teaching scientific medicine and encouraging clinical research, while proprietary schools continued to graduate poorly trained students who had little familiarity with new advances in medicine. The graduates of the new medical curriculum were viewed by the public as leaders in the community; physicians completing proprietary schools often reached lower social and economic levels (Starr, 1982).

These changes in medical education had been under way for twenty years when the 1910 Flexner report summarized the concerns of the American Medical Association Council on Medical Education. Flexner surveyed the science and clinical training offered by the U.S. medical colleges and reported that only a few in fact provided good-quality medical education. Many violated their own published entrance requirements and accepted anyone able to pay the tuition. Faculty, laboratories, and hospitals for clinical work were often inadequate. This public exposure of their failings encouraged the closure or merger of many of the inferior schools, so that the number of medical colleges declined from 155 in 1900 to 85 by 1920 (Kaufman, 1960).

The Flexner report prompted all medical schools to require entering students to have completed at least one year of undergraduate study in biology, chemistry, and physics (Flexner, 1910). A survey in 1914 found that 84 of 101 medical schools had in fact adopted this minimum entrance requirement, and 22 percent of the medical school graduates in that year held college bachelor's degrees (Brieger, 1983).

The distinction between basic research and clinical work in the modern medical schools gradually became more evident at this time. The research scientists were rarely involved in the care of patients, and the clinicians usually had little time for or experience in laboratory research. L. J. Henderson described the evolution of these branches of the medical faculties:

About the beginning of the present century faculties of medicine were often divided into two more or less hostile parties: the clinicians and the "laboratory men." Then, before long, it became clear that the clinical party had been defeated. Some of the reasons for this defeat were good reasons. Improvements in the methods of diagnosis and treatment called for new methods of medical education, because without scientific training the improvements were often incomprehensible. For the rising elite of the profession the experience of experimental research was often necessary, at least as a basis for self-criticism and for the appraisal of the work of others; perhaps still more as a means of achieving a feeling of equality with the laboratory men on their own ground, and so winning a sense of independence. Moreover, a man must be immersed in his medium in order to know it and acquire that comfortable sureness of action that is necessary for skillful action. And by this time the medium in which the physician worked had ceased to be merely sick men and women and had come to include all sorts of chemical, physical, physiological and bacteriological apparatus and processes. (Kunitz, 1983)

A prime example of this disparity involved the shift in emphasis within the Department of Medicine at Johns Hopkins which occurred under successive clinical leaders. William Osler headed the department until 1905, when Lewellys Barker assumed the position. R. Cole, who worked with both men, reported the changes:

Under Dr. Osler the opportunities for careful observations were never better and the importance of careful study of the more superficial aspects of disease never more insisted upon. But there existed in the clinic no great incentive to learn more about the fundamental nature of disease, and facilities for making the necessary investigations were lacking. The belief was general that medicine was basically different from biology or chemistry or anatomy and could only be studied by different methods. From time to time doubts about this point of view were expressed but chiefly by workers in the underlying sciences, and they usually held the opinion that the real investigations must be carried out by workers in their laboratories, since the clinicians had neither the time nor did they have the adequate training for these more complicated techniques. Dr. Barker on the other hand held that a primary function of a university department of medicine should be the encouragement of research and accordingly that the professor of medicine should be freed from the burdens of a private practice and allowed to devote his time to his own investigations and those of his staff. (Kunitz, 1983)

The parallel development of clinical and laboratory faculties in the modern medical school set the stage for the incorporation of the new science of human heredity into the curriculum.

Specific courses in human genetics and eugenics were offered in many colleges and universities throughout the United States within a few years of the rediscovery of Mendel's work in 1900. A survey in 1914 reported forty-four institutions with such courses (Davidson and Childs, 1987), among them Harvard, Columbia, Cornell, Brown, Wisconsin, and Pittsburgh (Haller, 1984). Three years later a total of seventy-three institutions offered genetics courses (Barker, 1917). Faculty and students at some universities established eugenics clubs to investigate the role of heredity in human reproduction and to educate the public on the importance of rearing healthy families. At the University of Wisconsin, several practicing physicians were involved in such a club, and medical students were invited to join (Baker, 1912).

But courses on human genetics were developed in the medical school curricula of only a few institutions. A suggestion was made in 1911 that the Hahnemann Medical College in Chicago establish a Department of Eugenics to emphasize the importance of human heredity as a means of improving the lot of the entire human race. The proposal was never implemented (Neiberger, 1911).

The first courses on human genetics for medical students were apparently offered at the Rush Medical College of the University of Chicago as part of the Division of Biological Sciences of the university. No prerequisites were required for medical students to enroll in any division elective courses. The medical school circular for the academic year 1912–13 described two lecture courses in the Department of Zoology: "Genetics—The problems of heredity in relation to genetics, to plant and animal breeding, and to eugenics;" and "Evolution and Heredity—A lecture course dealing with the evidences of organic evolution, human evolution, the history of the evolution idea, and its modern applications, the factors of racial descent, the physical basis and the laws of variation and heredity, modern experimental evolution, etc." Both courses were taught by William Tower (W. Kona, letter to the author, 1988).*

Another institution with a pioneering course on human heredity was the Washington University School of Medicine in St. Louis. Charles Danforth, an instructor there in the Department of Anatomy begin-

---

*Review of archival records at several other leading medical schools at this time has shown no evidence that courses in human genetics were part of the curriculum. Responses to inquiries at Harvard, Columbia, and Johns Hopkins were all negative in this regard (letters to the author from R. J. Wolfe, 1988, B. A. Paulson, 1988, and G. Shorb, 1989).

ning in 1908, had encountered "Mendelism" as an undergraduate at Tufts College in Boston and had also attended lectures on heredity by W. E. Castle at Harvard. In St. Louis he assisted in the medical school courses in anatomy, embryology, and neurology. Although doing research on the anatomy of fishes, Danforth attempted to learn as much as he could about human biology and during the summer of 1913 attended the course on eugenics sponsored by the Eugenic Record Office at Cold Spring Harbor.

During the next academic year he began to investigate the heredity of human malformations. In conjunction with the Ophthalmology Department of the medical school, he studied a large family with congenital cataract in three successive generations, results interpreted as evidence of a Mendelian dominant trait that was responsible for the defect segregating in the family (Danforth, 1914). Danforth shared his enthusiasm by organizing a course in human genetics for the medical students. Several reports indicate that this course was first offered in 1914 (Eugenics in the colleges, 1914; Willier, 1974), but the university archives report its first listing in the 1919 catalogue, a series of elective lectures described as "Variation and Heredity—The variations commonly met with in the study of human anatomy and their significance with special reference to heredity" (P. G. Anderson, letters to the author, 1988, 1991).

Despite these promising educational efforts, the formal recognition of human genetics as an important part of the medical school curriculum was not to come until the 1930s, when genetics courses were first required for medical students in the United States (Herndon, 1956).

These limited educational opportunities meant that few young physicians of the day had any formal exposure to human genetics during their medical school years. Their awareness of the rapid developments in this new science came from textbooks, medical journal articles, and discussions at medical society meetings by the leaders in the field of human genetics. These communications from the scientists set the tone for practicing physicians in their attempts to learn more about how heredity worked in human beings. But the perceived importance of genetics for the clinician was beginning to change, as reflected in the introductions to the two editions of Osler's *Modern Medicine*. The first, in 1907, noted enthusiastically, "Our views on heredity have been profoundly modified by the studies of Weismann, Mendel and others" (Osler, 1907). Only a few years later, the next edition stated, "Professor Adami's contribution on 'Inheritance and Disease' [has] been omitted.

Though important, [this] subject [has] not undergone very radical changes, and can be referred to in the first edition" (Osler and McCrae, 1913). The importance of human heredity for the physician in his daily tasks of diagnosing and treating important disease was increasingly questioned by many in the medical profession.

RESEARCH IN HUMAN GENETICS: THE IMPACT OF
THE EUGENICS RECORD OFFICE

Research programs in genetics were expanded at this time to augment the didactic courses already described at major universities such as Columbia and Harvard. The journal *Genetics* was established in 1916 to publish U.S. developments in this new scientific field. Philanthropists came to recognize the importance of genetics for agriculture and supported this work by endowing professorial chairs at colleges and universities around the United States. By 1915 genetics had become an academic entity with a secure professional position within the university hierarchy (Sapp, 1983).

Research on human heredity was pursued by both biologists and physicians. In 1914 F. A. Woods reported a summary of work in progress in this area (table 5.1). Some of the earliest work on human genetics in the United States originated from the Bussey Institute at Harvard (see chapter 3), where collaborative research between physicians and biologists was also performed. J. C. Phillips had been trained as a surgeon but possessed independent means and never practiced medicine. As a "voluntary assistant without candidacy for a degree," he worked with W. E. Castle on an investigation of the stability of the germplasm in experimental animals. The ovaries from female albino guinea pigs were surgically removed and replaced by the ovaries from a black strain of animals. Mated with albino males, these subjects produced only black progeny, demonstrating the stability of the black germplasm despite the altered environment after transplantation, and the dominance of black over albino in this species (Castle and Phillips, 1909; Castle, 1951). The success of this and other research involving animal subjects established the direction of future work when the administration of the Bussey Institute was reorganized in 1909: animal and plant genetics would be emphasized (Russell, 1954).

Lucien Howe, another physician who organized his own independent research programs in genetics, in 1876 became professor of ophthalmology at the University of Buffalo Medical School and surgeon-

Table 5.1.   Research in Human Heredity, 1914

Subject	Investigator
Acquirements	C. L. Redfield, M.D.
Consanguinity	H. H. Sturges
Deafness	Alexander Graham Bell
Epilepsy	F. A. Woods, M.D.
Feeblemindedness	H. H. Goddard
Great men	F. A. Woods, M.D.
Hare lip	W. F. Blades, D.D.S.
Huntington chorea	E. B. Muncey, M.D.
Inebriation	T. D. Crothers, M.D.
Insanity	A. J. Rosanoff, M.D.
Left handedness	H. E. Jordan, M.D.
Mental traits	C. B. Davenport

Source: Data from Woods, Meyer, and Davenport, 1914.

in-chief at the Buffalo Eye and Ear Infirmary (Kaufman, Galishoff, and Savitt, 1984; L. Sentz, letter to the author, 1991). He became interested in heredity as he investigated defects of the ocular muscles, observing that it was not unusual to find "three, four or more persons in the same family with similar forms of heterophoria, heterotropia, predisposition to ocular fatigue or similar abnormal muscular conditions. It seemed impossible to study these anomalies satisfactorily without first halting to learn something about that mystery which we call heredity" (Howe, 1919a).

Howe studied the inheritance of eye defects in several experimental animals during this decade. In the *Reliable Poultry Journal* he advertised for chickens with abnormal ocular development and obtained more than a dozen specimens that involved anomalies of the cornea and variations in the color of the iris, in the position of the pupil, and of the eyes themselves. T. H. Morgan sent him a blind strain of the fruit fly *Drosophila*. Howe was able to investigate the segregation of this character through more than a dozen generations of the insect. He performed these breeding studies at his summer home near Buffalo, which he called Mendel Farm (Howe, 1919a).

Toward the end of his long career, Howe relocated to Harvard Medical School and endowed a laboratory to investigate many aspects of ocular disease. He collaborated with the Eugenics Record Office staff to complete a *Bibliography of Hereditary Eye Defects,* eventually published in 1928.

The establishment of the Eugenics Record Office at Cold Spring

Harbor accelerated the development of human genetics as a science in the United States after 1910. The ERO had soon become the premier research institution for the study of human genetics in the country, and its director, C. B. Davenport, not only guided the ERO's research but within the decade became the leading speaker on issues regarding the interaction of genetics, disease, and society. He believed that many human traits were determined by single unit characters in the germ-plasm, a position outlined in the introduction to *Heredity in Relation to Eugenics:* "All men are created bound by their protoplasmic makeup and unequal in their powers and responsibilities" (Davenport, 1911). If human physical and moral qualities therefore could be brought under thoughtful biologic control, this thinking went, the betterment of society at large would certainly follow (Rosenberg, 1961).

The stated goals of the ERO were to foster human genetic research and then to apply findings to the public sphere so that people would choose their mates carefully. Davenport was convinced that human genetics should function not merely as a field for scientific investigation but as a basis for educational programs that would convince the public of the importance of eugenics for the improvement of human-kind in general (Allen, 1986).

Davenport recognized the intrinsic difficulties involved in the study of human genetics. Families tended to be small, generation times were long, and selective breeding was not possible. He was convinced, however, that the basic laws of heredity applied to humans as well as to other animals. Human genetic research could be done by collecting pedigrees and comparing patterns of inheritance with those already known from work in other mammalian species. The scientist should then be able to detect patterns of inheritance characteristic of dominant, recessive, and sex-linked Mendelian characters. Davenport recognized that the collection of such family histories would be an arduous task and should be undertaken by workers carefully trained in modern genetics (Punnett, 1912).

The ERO established summer programs to train the necessary field-workers in techniques for the collection of human pedigrees, and by 1917 some 156 students had completed the course of study. Assigned to investigate families in which an interesting hereditary trait had been reported, they examined family records and talked with family members, neighbors, and local physicians to gather medical and social information about the generations. They may have reported what Davenport expected them to find, for his belief in hereditary determinism was evident even in the training of the field-workers.

Because they were assigned to investigate traits believed to have a genetic basis, any environmental influence on social or medical traits was essentially discounted (Allen, 1986).

Davenport's espoused opinion that the goal of improving the heredity of the human race was both important and feasible was shared by other biologists and political leaders of the day (Ludmerer, 1972), but the biologists who studied plant and animal genetics demonstrated little interest in human heredity. The lack of any counteropinion from the biologic community at large implied that the claims of the ERO in the realm of human genetics constituted the best scientific knowledge available in that field of study (Allen, 1983). From the viewpoint of the general public, eugenics was human genetics.

From its inception, physicians were involved in the functioning of the ERO. Davenport invited several prominent physicians to become members of the Board of Scientific Directors: E. E. Southard, a neuropathologist from Harvard Medical School who also served as editor of the *Journal of Nervous and Mental Disease* and was a strong supporter of the country's eugenics movement (Kaufman, Galishoff, and Savitt, 1984); and William Welch and Lewellys Barker, from Johns Hopkins Medical School. But the involvement and influence of individual physicians on the actions of the board remains problematic, as the records of their meetings have been destroyed (Allen, 1986). Two major biographies of Welch fail to report anything about his service at the ERO, nor is the ERO included in the list of social and professional organizations of which he was a member (Flexner and Flexner, 1941; Fleming, 1954). Likewise the biography of Barker does not describe his work on the board of the ERO (Barker, 1942). By the time these books were written, the eugenics movement in the United States had acquired a rather unpopular social status, so the biographers may have chosen to selectively ignore these experiences in the lives of their subjects.

It is also possible that the inclusion of such prominent physicians on the ERO board was an attempt by Davenport to legitimize his scientific programs while actually allowing the men only minimal involvement in the operation of the organization. Correspondence between Welch and Davenport supports this contention. When Welch was first asked to serve at the ERO, he replied: "If you insist, I shall not decline, but I really feel that I should not go on any more national commissions or committees, and furthermore that you can find someone who will be better qualified and more helpful to you than I would be. Will you not consider putting someone else in my place?" He suggested that Barker be asked to serve. "He would carry weight by his reputation and position, is extremely interested in the subject of eugenics, has written about it and talks about it, is informed and withal a very able and wise man" (Davenport, 1936).

Davenport and Barker had served on the faculty at the University of Chicago in the early 1900s and were familiar with each other's interests. Both Welch and Barker eventually joined the ERO board. The Barker papers at Johns Hopkins contain no correspondence between Barker and the ERO (A. D. Ravanbaksh, letter to the author, 1992), but further statements from Welch reflect his position in the functioning of the board of directors. During 1913 he had been ill and very busy professionally. When Davenport suggested a board meeting, Welch replied: "I can hardly say when I should be able to attend a meeting. I did not know that I was acting as chairman of the Board. I really cannot assume even temporarily the responsibility of the chairmanship . . . Will you not proceed, as you may deem best, as regards the call for a meeting of the Board . . . You have these matters so fully in hand that I am quite ready to be guided by your good judgment . . . I feel that I am a very useless member of your Board, but I am interested" (Davenport, 1936). Davenport's leadership is clearly reflected in this correspondence. While the physicians on the board functioned as figureheads, his opinions set the course for the future direction of work at the ERO.

Other physicians became involved in human genetic studies, and at least seven completed the course for the training of field-workers and were qualified to collect pedigree data for the ERO (Allen, 1986). One of these was Elizabeth Muncey, who studied the inheritance of Huntington chorea and established with Davenport that it segregated as a dominant trait (Davenport and Muncey, 1916). The apparent success of the ERO in collecting genetic information and applying it to specific human conditions convinced many people that this organization constituted the best source of knowledge to guide the public toward social improvement (Allen, 1986).

Davenport appeared at numerous medical meetings during this decade to discuss the important role of human genetics in medicine and society at large. The American Academy of Medicine in 1909 provided a forum to discuss the role of heredity in human diseases. Davenport assured his physician audience that many human conditions were hereditary, such as albinism, retinitis pigmentosa, chorea, and imbecility. In other families a liability to a particular class of disease seemed to be inherited: throat, ear, skin, and lung conditions tended to run in certain families. He stated that environment certainly was important in causing human disease, "but unfavorable environment collects its toll first from those who are by heredity least resistant" (Davenport, 1910b).

At a 1912 meeting of the American Medical Association Davenport argued that all kinds of neurologic disease had a genetic component. This knowledge should not be kept private, he felt, but must become widely known to prevent the mating of unfit individuals. In this regard he viewed eugenics as a tool for "state sanitation" (Davenport, 1912a). His research on the heredity of social characteristics was featured at the 1913 meeting of the Chicago Medical Society, where he interpreted ERO pedigrees to show that "violent tempers, threats of suicide, love of fabrication, impulses to steal and set fires to buildings, profane and obscene speech, and great indolence" all had a marked hereditary basis. "Feeblemindedness" and kleptomania appeared to be inherited as recessive traits, while eroticism and violent temper were clearly dominant characters in many families (Davenport, 1913).

Davenport's hereditarian theme reached an even wider audience as he discussed the importance of sound genetic policy at the International Congress on Hygiene. In his view:

social progress is largely, if not chiefly, due to socially proper and fecund matings;
social decline is largely, if not chiefly, due to socially undesirable fecund matings;
permanent social improvement is got only by better breeding.

He argued that the public was deceived in thinking that improved nurture could ever overcome the effects of "deficiencies of breeding" (Davenport, 1912c).

Such authoritative statements were reinforced by publications in medical journals of the research sponsored by the ERO. As the decade progressed, these reports focused increasingly on the inheritance of social and mental traits. For example, in the field-workers' investigation of 175 families with epilepsy, the inheritance of the character was interpreted to show that it acted as a single unit character (Davenport and Weeks, 1911). Imbecility appeared to segregate as a dominant trait (Davenport, 1910–11). A review of the family history from 165 "wayward girls" showed that their violent tempers had also been inherited as a dominant trait (Davenport, 1915a). And generalized "neuropathy" was a trait that appeared to be inherited in a recessive manner (Rosanoff, 1912).

Parallel reports in leading scientific publications further legitimized the efforts of the ERO staff. The neuropathic trait was defined as "insane, epileptic, hysterical, feebleminded or unusual conduct." Review of the data from seventy-two families showed that this trait segregated as a recessive character (Rosanoff and Orr, 1911a, b). Violent temper and uncontrolled eroticism appeared to be dominant traits, while wanderlust was often inherited as a sex-linked character (Davenport, 1914). Insanity often segregated as a recessive trait (Rosanoff and Martin, 1915), while chorea clearly fit a dominant pattern for inheritance (Davenport, 1915b, c). The evidence on "feeblemindedness" was not quite so straightforward. Davenport believed that the trait was most likely a result of a group of unit characters often, but not always, inherited together. The major conclusion of the study was that much "feeblemindedness" was in fact hereditary. That finding alone was sufficient to justify a eugenic campaign, Davenport thought, while awaiting further research on the technical aspects of its behavior (Feeblemindedness, 1915a).

The work of the ERO during this decade was interpreted to mean that most human characteristics—mental, physical, and social—were controlled by single unit characters inherited from the parents. This rigidly hereditarian philosophy was summarized by Davenport at the International Congress of Eugenics in 1921, where he presented family data on many human conditions that his work had shown were caused by heredity: deafness, skin color, hemophilia, "feeblemindedness," epilepsy, dementia praecox (schizophrenia), depression, and skill in art, music, and sailing all were the result of heredity. His hope was that society would recognize the importance of heredity for both good and evil character traits. The future of human society, in his opinion, depended on the preservation of the basic sound American stock (Davenport, 1921).

GENETICS AND CLINICAL MEDICINE: THREE LINES OF THOUGHT

"Why not use even a small part of the energy spent on theoretical discussions to demonstrate that the thing is actually possible? . . . it is accessible to experimental verification." Thus did T. H. Morgan summarize the aim of biologists during this period: to test all hypotheses experimentally (Allen, 1968). He was discussing the alleged heredity of acquired features in animals and plants, but the goal of hard science was evident throughout the genetics community. The controversy within the medical profession about the existence of the heredity of acquired characters continued during the first two decades of the new century and encapsulates much of the tension between traditional and contemporary opinions on how heredity functioned in human beings. When two siblings were born with microtia (small, malformed ears), the question of maternal impression during the pregnancies was voiced as late as 1904 as a plausible explanation for the defect in embryonic development (Braishin, 1904), even though most biologists agreed that such parental control of fetal development and the heredity of influence (telegony) were unsubstantiated by scientific evidence (Jordan, 1914; Moleen, 1919). That many people still believed in these phenomena, despite the claims of modern science, suggested to some that perhaps there existed a "deep psychological need" to believe that defects could result from such mechanisms (Kroeber, 1916).

With the development of the unified theory of genetics after 1910 came general agreement among biologists that the only conditions that could become hereditary were those that resulted in a change within the germplasm of the egg or sperm (Irwell, 1912; Moleen, 1919). But in the medical community a belief in the heredity of acquired features continued during this time. A seemingly endless series of articles by Redfield in the medical literature attempted to provide scientific evidence from work on racehorses to convince physicians that traits obtained by practice and training could be transmitted to offspring (Redfield, 1916a, b, 1917). Many other physicians argued instead that the available evidence weighed against the heredity of acquired features (Mack, 1911).

The ongoing controversy between traditional and more modern conceptions on the mechanism for heredity was evident in the three distinct lines of thought that developed within the medical community after 1910. Some physicians were satisfied with the explanation for heredity which had worked for them over the years and were not interested in the new genetics. A second group had heard about these new developments in genetics and tried to assess their applicability for human conditions. They argued that there *might* be something important developing in this field of research. A small third group slowly grasped some of the implications of this new genetics as a significant explanation for the causes of human disease. The second decade of the twentieth century was marked by intense controversy among these three groups of physicians.

Traditional Heredity

The nineteenth-century model for heredity, "Like begets like," continued to exert a powerful influence over traditional thinkers among U.S.
physicians well into the new century (Mack, 1911). Evidence for the
direct transmission of a character from parent to child still constituted
the most important evidence for heredity at work. Of migraine headache, for example, reported in successive generations of many families,
one physician observed that there was "no nervous functional disease . . . in which direct heredity plays so important a role" (Andvist,
1913). Twenty-five pedigrees with hemorrhagic telangiectasia (spontaneous bleeding from abnormal capillaries close to the skin's surface)
showed direct transmission through four generations. Therefore, "heredity is the only important etiologic factor" (Steiner, 1917). The observation of spastic paraplegia (a form of cerebral palsy) in three successive generations of another family indicated the "true hereditary
character of this type of degeneration of the central nervous system"
(Mason and Rienhoff, 1920). Parents and children in several other
families were found to have unusually fragile red blood cells, a familial
pattern that also suggested hereditary factors at work (Griffin and
Sanford, 1919).

Conversely, the absence of affected ancestors was viewed as strong
evidence against the hereditary nature of other specific traits. Wilson
disease (a degenerative disorder of the liver) was observed in several
brothers and sisters, but parents and other relatives were unaffected.
The disorder was certainly familial, but not hereditary (Hamilton,
1916; Spiller, 1916). Gaucher disease (another degenerative disorder
involving the liver and central nervous system) was observed in a similar pattern of affected siblings, likewise in families with no other evidence of the disorder. This was believed to demonstrate that the character was not hereditary in nature (Reuben, 1914, 1918). The same
family pattern in peroneal nerve atrophy and several bone disorders
was interpreted to imply lack of support for heredity as a factor in
causing them (Hatch, 1915; Young, 1916).

The apparent familial nature of other traits did not make sense in
terms of like begetting like. Hemophilia ran in families, apparently
passed through successive generations by unaffected females who produced affected male offspring. But its "curious disposition to transmission through the females remains wholly without explanation" (Camac, 1915). This familial pattern of disease was attributed to the "law of
Nasse" from 1820 (Burch, 1912; Hess, 1916). The etiology of amauro-

tic family idiocy was likewise called "mysterious." Why should only one or two children in a family of seven be afflicted? (Epstein, 1917a).

In situations where heredity did appear to play a role in causing specific familial diseases, the agent transmitted was still generally believed to be a predisposition to the disease and not the disease itself. This mechanism of diathesis, which produced symptoms of disease after an appropriate environmental trigger, was accepted as an adequate explanation for the pathology in periodic paralysis (Atwood, 1912), neurofibromatosis (von Recklinghausen disease) (Barss, 1917), St. Vitus dance (Burr, 1917), hair-nail dystrophy (Eisenstaedt, 1913), and epilepsy (Burr, 1922).

Such a model for the cause of diseases seemed to fit the available facts, and therefore no revision of the existing paradigm on human heredity appeared to be necessary for many physicians. In the case of muscular dystrophy, Camp stated emphatically that this disorder could not be a hereditary trait that had been transmitted as a Mendelian unit character. Instead, a degenerative factor had been directly transmitted in such families (Camp, 1916). As Epstein put it, certain children were born with "wrecked constitutions." When exposed to external factors, a myriad of diseases could develop (Epstein, 1917b).

The importance of unique susceptibilities to disease was evident even in the case of infections. Not all exposed people actually developed symptoms of such diseases. Certainly a hereditary "pathologic disposition" must exist which could explain the differences in the frequency of diseases between different individuals (Hymanson, 1918). This predisposing factor was the thing actually inherited. The production of disease symptoms almost always required an external trigger, which acted on susceptible individual tissues and organs. This model for hereditary disease was accepted by some physicians as it had been since 1860. But did it constitute a scientific explanation acceptable to modern medical scientists?

Rumors of Mendelism

Physicians of this era had acquired more training in the biologic sciences than had their predecessors, and many were coming to view medicine as more a science than an art or a craft. While this group did not reject the developments in the new science of genetics, their relevance for routine medical work remained obscure for most. These physicians could discuss the unified model for heredity involving the segregation of unit characters carried by the chromosomes during cell division (Connolly, 1912; Hichney, 1912), but they appeared to have great difficulty in seeing how such mechanisms could explain the familial patterns of human heredity in specific cases.

For example, in a report of the same dental anomaly (absent superior laterals) in three lines of a family, suggesting that a hereditary factor was involved in the altered embryogenesis, the author discussed examples of Mendelian segregation in plants and animals but could not interpret his own data in genetic terms (Stanton, 1914). The author of another report describing fourteen eye disorders observed in several generations of particular families thought that Mendel's laws controlled the transmission of human physical defects, but their application to these cases remained an enigma (Fischer, 1914). Other physicians made similar comments regarding heredity and migraine headaches and a variety of other neurological diseases (Burnett, 1912; Gardner, 1913–14; Buchanan, 1920). The Mendelian model for heredity was described repeatedly in general terms, but these physicians appeared unable and often reluctant to apply the new model to specific cases of human disease.

Such physicians sensed something important in these new notions on heredity. Of a family with the hand malformation hypodactylism, it was noted that "the Mendelian law may apply in this case" (Cowie, 1913). The cause of amaurotic family idiocy remained elusive. One physician expressed the hope that "the Mendelian principle may perhaps supply the missing etiology" (Epstein, 1920). But physicians were often unsure as to the importance of such hereditary factors in the causation of human disease. In a large family with diabetes in several different lines, the question was raised whether "the diabetic predisposition behaved as a Mendelian unit character" (Allen and Mitchell, 1920). The appearance of congenital cataract in many individuals of another family forced one physician to admit that "Mendel's laws and suggestions must command our attention and add to our wonder" (Campbell, 1913). Even at the end of the decade, a study of migraine headache in 127 families led Buchanan to conclude that it was a "Mendelian phenomenon, " but he was unable to categorize its pattern of inheritance in any greater detail (Buchanan, 1920).

The nagging question for the physician of the day was whether Mendel's laws could be applied to human heredity in the first place. In the case of familial aniridia (absence of the iris), Risley believed that the state of knowledge regarding human heredity was quite limited and "any effort to explain the essential nature of aniridia would . . . lead to unprofitable speculation . . . on the mechanical explanation of Mendel's laws of heredity" (Risley, 1915b). Likewise it remained question-able whether Mendelian ratios were found in human families (Howe, 1918, 1919a; Buchanan, 1921).

The New Genetics and Medicine

The Eugenics Record Office was viewed by many U.S. physicians as the primary source of information on human heredity at this time. Daven-port and his co-workers collaborated with physicians and dentists on several research projects that investigated the genetic basis of specific human diseases during this decade, among them Blades, a dentist who used the pedigree records from the ERO to assess the importance of maternal impressions in causing cleft lip and palate. He expected that the results of such an inquiry would then be used to guide members of affected families in marriage choices (Blades, 1914). The ophthalmolo-gist Howe also worked with Davenport and examined records from twenty-four families with eye diseases. He found evidence for reces-sive, dominant, and sex-linked patterns of inheritance for specific eye defects (Howe, 1918). Davenport was also asked to review a large ped-igree with congenital cataract prepared by Ziegler. In Davenport's opinion, the trait segregated as a dominant character, being transmit-ted to about half the children born of an affected parent (Ziegler, 1915). Preiser and Davenport studied twenty-seven families with neu-rofibromatosis (von Recklinghausen disease), which also segregated in a dominant pattern, tending to reappear in successive generations (Davenport, 1918; Preiser and Davenport, 1918).

Heredity in Relation to Eugenics (Davenport, 1911) was the primary written source on human heredity used by U.S. physicians during this era.* It was cited in numerous reports on families with disorders as varied as coloboma irides (malformed iris) (Lewis, 1915), polydactyly (Arps, 1920), osteogenesis imperfecta (soft, fragile bones) (Terry, 1918), and epilepsy (Kehoe, 1915). In each instance the pattern of heredity described by Davenport was accepted by the physicians as fact, whether or not the explanation fit the observed data. In certain instances, he predicted that *all* progeny of one parent affected with a dominant trait could be afflicted with the same disorder, although the expected Men-

---

*The other genetics texts in print at this time were written by British biologists: Thomson (1908); Bateson (1909); and Punnett (1911). They were virtually unknown to U.S. physicians and thus exerted no influence on their thinking about heredity and disease.

delian ratio in such cases was actually one-half. Such statements were clearly motivated by eugenic considerations. Davenport expected that his collaboration with physicians would result in their wisely guiding prospective parents into appropriate marriage choices.

Other physicians used the ERO model for research as the basis for their own investigations of human heredity. Weeks trained several field-workers, who obtained family histories in 397 cases of residents at the New Jersey Village for Epileptics. He assumed that epilepsy segregated as a unit character and analyzed the parental crosses and offspring affected with various neurologic diseases (Weeks, 1912). Boyd reported a family with hereditary chorea and appended a "chart of the family along Mendelian lines" (Boyd, 1913).

About this time in a number of physicians' minds the Mendelian patterns of segregation of unit characters began to make sense when applied to human pedigrees involving many different traits. William White was among the first U.S. physicians to explain the familial occurrence of disease in Mendelian terms. The superintendent at St. Elizabeth's Government Hospital for the Insane in Washington, D.C., he also served as professor of nervous and mental disease at both Georgetown and George Washington University medical schools (White, 1938). White believed that it was important for modern physicians to know something about the new science of genetics, previously a subject for investigation by botanists and zoologists. But now physicians were becoming interested in the field because of "the problem of eugenics— the preservation and improvement of the race" (White, 1909).*

As early as 1911 White requested that the ERO send a field-worker to Washington to investigate familial disease that he observed in clinical practice, which he believed would prove a fertile field for study. He noted that many hospital patients came from large families in which a variety of neurologic diseases had been reported (White, 1911).

White's 1913 text, *Modern Treatment of Nervous and Mental Disease*, explained the familial pattern of many disease traits in terms of Mendelian inheritance. The regular parent-to-child inheritance of Huntington chorea and familial tremor fit a dominant mechanism. The appearance of Friedreich ataxia in several brothers and sisters, even though the parents and other relatives were unaffected, fit the Men-

---

*Whether White shared his eugenic concerns with medical students is not apparent. A review of the catalogues before 1920 at Georgetown University School of Medicine mentions nothing about genetics in the courses which he taught (B. King, letter to the author, 1991).

Table 5.2.   Human Genetics Studies Providing Pedigrees, 1911–1922

Trait	Reference
Direct heredity	
Manic depression	McGaffin, 1911
Huntington chorea	Ballentine, 1912
Hair-nail dystrophy	Eisenstaedt, 1913
Psoriasis	Engman, 1913
Glaucoma	Calhoun, 1914
Chondro-osteoma	Percy, 1915
Hereditary exostoses	Montgomery, 1916
	Maynard and Scott, 1921
Cataract	Parker, 1916a, b
Myotonia congenita	Toomey, 1916
Deforming chondrodysplasia	Ehrenfield, 1917
Hemorrhagic telangiectasia	Steiner, 1917
Erb muscular dystrophy	Timme, 1917
Polydactyly	Arps, 1920
Mendelian heredity	
Migraine headache	Burnett, 1912
Epilepsy	Weeks, 1912
Huntington chorea	Boyd, 1913
Anodontia	Stanton, 1914
Mendelian dominant heredity	
Huntington chorea	White, 1913
	McKinnis, 1914
Familial tremor	White, 1913
Cataract	Danforth, 1914
	Ziegler, 1915
	Buxton, 1916
Polydactyly	Brandeis, 1915
Ectopia lentis	Buxton, 1916
Hereditary ankyloses	Cushing, 1916a
Angioneurotic edema	Crowder and Crowder, 1917
	Phillips, 1922
Ptosis	Briggs, 1918
Mendelian recessive heredity	
Friedreich ataxia	White, 1913
Amaurotic family idiocy	Herrman, 1915a, b, c
	Brandeis, 1918
Mongolian idiocy (Down syndrome)	Herrman, 1917
Mendelian sex-linked recessive heredity	
Duchenne muscular dystrophy	White, 1913
Unclear heredity	
Feeblemindedness	Goddard, 1912
Amaurotic family idiocy	Coriat, 1913
Aniridia (Mendelian?)	Risley, 1915b
Familial tremor (direct?)	Craig, 1916

Table 5.3   Familial Disease and Mendelian Inheritance, 1910–1921

Trait	Pedigree Provided in Study?	Reference
Mendelian recessive heredity		
Friedreich ataxia	Yes	White, 1913
Amaurotic family idiocy	Yes	White, 1913
	Yes	Herrman, 1915a
	Yes	Herrman, 1915b
	Yes	Herrman, 1915c
	Yes	Brandeis, 1918
Unclear heredity		
Amaurotic family idiocy	No	Abt, 1911
	No	Smith, 1912
	Yes	Coriat, 1913
	No	Hymanson, 1913
	No	Sachs, 1915
	No	Epstein, 1917a
	No	Epstein, 1920
	No	Welt-Kakels, 1917
Friedreich ataxia	No	Spiller, 1915a
	No	Lloyd and Newcomer, 1921

delian inheritance pattern for a recessive trait. And the special case of muscular dystrophy with affected males born of unaffected females could be explained as a sex-linked recessive character (White, 1913).

The pedigree, previously a symbol for direct parent-to-child heredity of some predisposing trait, during this decade came to mean something quite different as a gradual transition took place in the U.S. medical literature on genetics. Although the pedigree served its traditional purpose, it came to represent a Mendelian interpretation of much of the family data (table 5.2). As the decade progressed, characteristic patterns of familial disease represented by the pedigree were interpreted as evidence for Mendelian dominant, recessive, and sex-linked modes of inheritance.

Even disorders that affected several siblings with no affected antecedent relatives began to be explained in terms of the segregation of recessive unit characters, as evidenced by reports on Friedreich ataxia and amaurotic family idiocy which appeared during this decade (table 5.3). These disorders had been labeled familial but not hereditary in the traditional sense because no one else in the family had been previously afflicted. Now these familial occurrences could be represented by

Table 5.4. Selected Human Mendelian Traits, 1910–1920

Trait	Reference
Dominant	
Huntington chorea	White, 1913
	McKinnis, 1914
Tremor	White, 1913
Cataract	Danforth, 1914
	Buxton, 1916
Polydactyly	Brandeis, 1915
	Gordon, 1915
Ectopia lentis	Buxton, 1916
Symphalangism	Cushing, 1916a, b
Angioneurotic edema	Crowder and Crowder, 1917
Ptosis	Briggs, 1918
Osteogenesis imperfecta	Terry, 1918
Recessive	
Cancer	Levin, 1912
Friedreich ataxia	White, 1913
Amaurotic family idiocy	Herrman, 1915a, b, c
	Brandeis, 1918
Epilepsy	Weeks, 1915
Spinal gliosis	Williams, 1915
Oppenheimer disease	Coriat, 1916
Mongolian idiocy (Down syndrome)	Herrman, 1917
Asthma	Adkinson, 1920
Sex-linked	
Muscular dystrophy	White, 1913
Hemophilia	Hess, 1916

a pedigree. The traits were indeed hereditary, with parents who were unaffected heterozygotes bearing homozygous recessive affected children. The familial pattern of many disorders was clarified in the light of the new Mendelian theory.

An impressive array of conditions that matched the expectations of Mendelian inheritance provided strong evidence for its utility in explaining these examples of human heredity (table 5.4). One physician was convinced that such information indicated the existence of a series of determinants within the germ cells which controlled the development of all the organs of the human body (Moleen, 1919).

The segregation of unit characters had first been substantiated in plant and animal breeding experiments. Observation of the inheritance patterns of specific human characters now seemed to parallel the Mendelian ratios observed in the other species. The correctness of the new model for heredity was given credence by its success in describing rules for the inheritance of disease traits, where its predictions often fit the observed results. Because it also worked better than the models for inheritance based on a blending of characters or diathesis, the unified model of unit characters segregating on particular chromosomes was widely accepted by many physicians as the right theory for human heredity at this time.

THE LIMITATIONS OF THE NEW GENETICS

In 1922 the biologist E. M. East reviewed the state of genetics in an article titled "As Genetics Comes of Age," in which he argued that the new science had been firmly established on experimental facts derived from the study of animals and plants. Genes appeared to be physical entities that followed the segregation of chromosomes from one generation to the next. The key concepts of independent segregation of characters, linkage groups, and the role of individual chromosomes as linkage groups all had been substantiated by work in cytology and experimental breeding (East, 1922).

The study of human heredity had not matured as rapidly as that of plant and animal genetics. It remained largely descriptive, not analytical, and several intrinsic difficulties hindered new developments. Human beings were not favorable experimental subjects. Matings of individuals with particular traits could not be controlled; generation times were long, with hereditary information rarely available for more than three generations; family size was small, and women frequently miscarried; and individuals who might have eventually demonstrated the hereditary character under investigation often died young of infection. All of these facts made accurate counting of unaffected and affected progeny inaccurate at best. The inability to be sure that the data were sound did nothing to convince either biologists or physicians that the calculated Mendelian ratios could be compared with the expectations of the theory (White, 1913; Ingham, 1915; Ludmerer, 1972).

Available laboratory techniques also had limitations when applied to human tissues. Attempts to identify and count human chromosomes, for example, had produced quite variable results. The processes of cellular death, chemical fixation, and staining could easily generate artifacts that did not accurately represent the status of the chromosomes *in vivo* (Hichney, 1912). Fixed gonadal tissue was the usual source of material for cytologic study, and workers from differ-

Table 5.5   Representative U.S. Studies of Human Diploid
Chromosome Number, 1900–1925

Tissue	Diploid Number	Reference
Testis	18	Wilcox, 1900
	24	Castle, 1909
	24	Wieman, 1917
	46	Painter, 1921
	48	Painter, 1922
	48	Painter, 1923
Gonad	22, male; 24, female	Guyer, 1910
Brain	33–34	Wieman, 1912
Liver	34	Wieman, 1912
Nose	34	Wieman, 1912
Skin	34	Wieman, 1912

ent laboratories reported human diploid chromosome numbers ranging from eight to more than fifty per cell (Hsu, 1979) (see table 5.5).
The work by Painter epitomizes the difficulties inherent even in efforts
using the best available techniques. His initial report found forty-six
chromosomes in each human cell. Further study caused him to revise
the figure upward to forty-eight. His work represented the apogee of
human cytogenetic study and was accepted as correct until the 1950s,
when new techniques of tissue culture and tissue preparation would
allow better separation and characterization of the human chromosomes. After 1923, laboratory work on human chromosome structure
and segregation in normal and diseased tissues essentially halted for
decades.

The quality of the available data on human genetics was a matter of
great concern for both physicians and biologists. If genetics was to be
successfully applied to humans for the purpose of social improvement,
the information needed to be correct. Blackford told a medical meeting that "it is highly important for the welfare of our race that the laws
regulating the transmission of physical, mental and moral traits should
be known thoroughly and accurately, for it is by such knowledge that we
will be able to proceed toward the elimination of hereditary weakness
and the strengthening of hereditary virtue" (Blackford, 1915–16).

Careful analysis of multiple generations of several families was
required to discern the pattern of inheritance for any particular trait.
Reports without such careful inquiry were worthless, and shoddy
methods of investigation could bring the entire science of genetics into
disrepute (Genealogy and eugenics, 1915). Because "slippery and inse-

cure foundations" were not acceptable in human genetics, workers in the field were cautioned about letting their "hopes run riot with the facts" (Kohs, 1915).

Much of the research in this area involved the analysis of large numbers of pedigrees such as those collected by the field-workers from the ERO, whose techniques raised many questions that could harm the integrity of human genetics as a science. Critical examination of their data revealed that, while information on people interviewed was fairly accurate, the data on noninterviewed persons often were not (Little, 1922). In a large number of pedigree reports no information was available for about one-half of the members in a family, making an analysis of the numbers of unaffected and affected individuals impossible (Heron, 1914). Conclusions on the Mendelian segregation patterns of particular human traits could not logically be drawn from such incomplete data (Heron, 1913).

White reported from Washington that "here in the Hospital we have had a fieldworker, educated at the Cold Spring Harbor Department of the Carnegie Foundation, and she has been working upon family histories and she has worked out many family histories of patients in this Hospital, and she has spent weeks and weeks upon these records, and I challenge you or any other person to point to one of these carefully worked out families and give me in a single instance a prediction of what the heredity of any one of the numerous individuals therein contained will be shown by our charts that it actually is" (White, 1919). His practical experience therefore did not convince him that this new genetics was at all useful in clinical medicine.

Furthermore, persons outside the ERO suspected that the fieldworkers were often told what to find before their investigations began (Kohs, 1915). One physician summarized this opinion: "The learned biologist may hope and imagine that the most sanguine expectation of the Mendelists may be realized, but the dabbler in science is positive that they will and talks as if they were already realized" (Timme, 1914).

A basic assumption of many of the research workers in human genetics was that most human traits were controlled by single unit characters in the germplasm (Kohs, 1915). Most physicians, on the other hand, believed that such characters were generally more complex and could not be inherited in such a simplistic manner. In fact, they believed that many human traits such as epilepsy, mental retardation, mental illness, and alcoholism were symptoms of other underlying defects. Before accurate studies on the inheritance of such traits could be undertaken, more precise diagnostic tools were required

(Davenport, 1912b). The imprecise application of genetic theory to poor-quality data could therefore be used to explain the heredity of "almost anything" (The feebly inhibited, 1915).

Another problem involved nonphysicians collecting family data on medical conditions. The field-workers from the ERO, for instance, were expected to ascertain the presence or absence of epilepsy or other neurologic diseases in family members without the technical ability to diagnose such conditions in any objective manner (Hecht, 1913). In fact, the recognition of this limitation may have fostered collaboration between the research biologist and the physician. For example, when H. H. Newman of the University of Chicago attempted to study the inheritance of nightblindness in members of a large family and could not accurately diagnose the condition, he suggested that the ophthalmologist and geneticist work together to study the inheritance of such diseases (Howe, 1918).

The application of Mendelian laws to human inheritance was certainly not as straightforward as it had earlier appeared. Walsh observed that "the question of heredity is a complex one, and only those fully acquainted with its various ramifications and apparent contradictions should presume to draw conclusions in any given case, and then only after a thorough investigation of the facts" (Walsh, 1918). Davenport, for example, having reported a study on the heredity of congenital cataract which implied that it segregated as a dominant trait, counseled against the marriage of individuals from such affected families. Re-evaluation of the same family pedigrees showed that a recessive pattern of inheritance was more likely. The criticism in this case involved not only bad science but also the misapplication of the bad science for eugenic purposes: "It is not only because of a mistake in the method of inheritance, but such rules should never be made until the exact hereditary processes are positively known, since such practices are likely not only to bring discredit upon the science, but to injure people who endeavor to follow them in the regulation of their lives" (Jones and Mason, 1916).

While genetic principles appeared to work well in experimental breeding studies involving plants and animals, Hecht and other biologists and physicians expressed doubt that they would "fit so snugly into a scheme to explain diverse human traits" (Hecht, 1913). Many examples were reported in which the laws did not appear to work as expected. In a family with four generations, hereditary aniridia (absence of the iris) was expected in about one-half of the progeny, under the assumption that it segregated as a dominant trait. In fact the family reportedly had 112 of 117 affected individuals, far more than predicted by any known theory of inheritance (Risley, 1915a). One physician observed that Mendel's laws did seem to hold for humans, "but as yet verification of the laws has not been carried so far, although we believe we are on the way to fuller light along this line" (Blackford, 1915–16). As current knowledge on the heredity of specific human traits was generally incomplete, its application might often produce unsatisfactory outcomes. Then the "exciting interest and confidence in the eugenics movement" could be dealt a "severe and possibly permanent setback" (Little, 1922).

CAUSES OF HUMAN DISEASE: HEREDITY VERSUS INFECTION

The ongoing controversy regarding the relative roles of heredity and environment in causing disease continued to confuse many physicians. A speaker at the St. Louis Medical Society commented that "the physician is prone to regard the entire subject as something extremely vague" (Hinchey, 1913). Biologists criticized physicians for showing interest in the heredity of normal traits such as eye and hair color, but not in pathologic features (Herrman, 1918). White agreed that this was the case. In his opinion, the labeling of a character as "hereditary" actually hindered its further study because it "created the delusion that everything has been explained" (White, 1913). If a trait was hereditary and therefore fixed, any treatment was useless.

The success of the germ theory in explaining many human diseases was reflected by public health measures, which achieved a significant diminution in their frequency. Because the bacteriologist appeared to hold the key for the comprehension of much human pathology, medical research in this era focused almost exclusively "in the field of germ and cellular pathology" (Hecht, 1913; Ransohoff, 1913). The professor of pathology at Harvard, W. J. Councilman, believed with most physicians of the day that human disease was produced by external factors acting upon individuals (Councilman, 1913).

The major thrust of medical research after 1910 explored how such environmental factors could alter normal human physiology and result in various disease states. A strong antihereditarian trend evolved as the causes of more and more human diseases were identified as germs or toxicants. Dercum believed that neurologic disease, for instance, resulted from altered physiology. "The facts suggest that we have to do with biochemical processes akin to those of infection and immunity" (Dercum, 1915).

The etiology of human cancer was another area of considerable debate at this time. In the past, heredity had been considered to have substantial influence in causing this malignancy (Bidwell, 1911). Analyzing data collected by the ERO on familial cancers, Levin found certain lineages in which between 10 and 33 percent of the offspring were affected, which he interpreted to "correspond very clearly to the Mendelian percentage of members with recessive unit-characters in a hybrid population." The type of cancer also bred true in different families. In one the uterus was affected, in others the breasts. This evidence convinced Levin that familial cancer resulted from "the union of two germplasms, each of which is characterized by the presence of germ cells that are non-resistant to cancer" (Levin, 1912).

Warthin reanalyzed the same data and agreed that the susceptible families demonstrated a risk for cancer twenty- to thirtyfold greater than that of the general population, but he was not sure that the data defined a Mendelian pattern for the inheritance of specific cancers (Warthin, 1914). When he then reviewed the family histories of 1,600 patients with cancer from the University of Michigan Hospital between 1895 and 1913, he found familial examples of specific tumors. Those commonly evident in many family members involved cancer of the lip, mouth, breast, stomach, uterus, and intestines. But none of the pedigrees demonstrated a Mendelian pattern for the inheritance of such cancer susceptibility (Warthin, 1913).

At this time Maude Slye began a series of studies on the inheritance of cancers in strains of inbred mice. She found that cancer susceptibility could be raised or lowered by selective breeding in these animals and that the inheritance of this characteristic appeared to follow Mendelian predictions. Resistance to carcinoma was a dominant trait, while susceptibility behaved as a recessive trait (Slye, 1915, 1916). Roundly criticized in the biologic literature (Little, 1915, 1916), her work was nevertheless seriously considered by the medical profession (Bristol, 1916). It encouraged further investigation of possible hereditary influences on the formation of human cancers. A review of life insurance records showed, however, that the presence of an affected parent or sibling did not increase the risk of cancer in other living family members (Is cancer either, 1917).

The available data on familial cancer also raised the question of the alleged infectious nature of human cancers, for the life insurance records provided no evidence that family members who cared for a patient with cancer were at any greater risk for developing the disease than those in the general population (Is cancer either, 1917). Although this argued against the role for an infectious agent in these familial cancer cases, other studies were interpreted to show that such familial disease was caused by a parasite, an infectious agent, just as were syphilis or tuberculosis (Smith, 1908; Abelmann, 1911), in which case personal hygiene and public health measures would be advised to decrease the likelihood of further spread in the community at large (Byford, 1915). Even if cancer was not an infection, it certainly was a result of chronic poisoning of the tissues by toxins resulting from faulty metabolism and elimination (Perdue, 1914; Bristol, 1916).

Cancer was believed to share many features with the important human disease tuberculosis. Both recurred in several generations of certain families, affecting several individuals in each generation. The tubercule bacillus had been identified as the causative agent for tuberculosis, and some evidence suggested that cancer was also an infectious disease, but not all members of such families developed symptoms of either disease. That individual susceptibility could easily be demonstrated suggested a hereditary component. The opinion of many physicians then was that any hereditary tendency to cancer formation probably involved a complex type of inheritance that was not strictly Mendelian in nature. Its relevance to clinical medicine, however, was believed to be "of negligible importance to a practical preventive measure in man" (Little, 1916). Cancer then appeared to be more a public health problem.

The same type of controversy existed regarding the causes of epilepsy. Some studies had shown the frequent occurrence of epilepsy in parents and their offspring, but a more detailed study of 175 pedigrees of families with epileptic persons discovered only twelve other relatives who had the disease. Thus heredity appeared to play a minor role in causing human epilepsy (Clark, 1912; Fairbanks, 1914). More important causes of epilepsy appeared to be trauma or infection involving the central nervous system (Hunt, 1911; Fairbanks, 1914). In fact, the isolation of a spore-forming bacterium from the blood of many epileptic subjects shortly after they had a seizure strengthened the opinion that convulsions could be triggered by an infection (Munson, 1917). The development of a seizure was therefore believed to represent irritability of the central nervous system by toxins, infections, or injuries.

Infection also seemed to cause other diseases that occasionally ran in families. Ichthyosis was viewed as a congenital infection in which microorganisms transmitted from mother to child produced altered skin function (Young, 1910–11). Epidermolysis bullosa—the spontaneous, widespread blistering of the skin—was also believed to be an infection that could be passed through several generations in particular families (Ravogli, 1917).

The accumulated evidence suggested to many physicians that heredity as an important cause of human disease had been overrated in the past and substantiated the role of infections as agents in diseases affecting millions of people. Antihereditarian opinions were expressed repeatedly in the medical literature: not only could further attention to human heredity divert the efforts of the medical community from the daily management of potentially treatable diseases, but the prospects heredity offered were often too gloomy. One physician called it a "bug-bear" (Wright, 1914). A writer of a letter to the editor of the *Journal of the American Medical Association* remarked that "heredity has been made the scapegoat for many ills actually due to defects in the present social organization" (Ager, 1919). Physicians were urged to stop blaming ancestors for causing diseases in their descendants. Rather, the community should attempt to improve the "work, play, diet and sleep" in the childrens' environment (Keogh, 1916).

Although the importance of infectious agents in causing much of human disease was recognized by U.S. physicians at this time, it remained clear that not everyone exposed to the germs developed clinical disease. There had to be constitutional differences in susceptibility among different individuals. Thus the traditional notion of predisposition to disease, possibly influenced by heredity, continued to be examined as it applied to both physical and mental traits in patient populations (Herrman, 1918; Hymanson, 1918).

THE EUGENICS DEBATE: ROLE OF THE ERO AND PHYSICIANS

Within the eugenics community, the opposite trend toward social Darwinism continued during this decade. The general opinion that most human characters were determined by hereditary factors provided a theoretical basis for attempting to control human reproduction with the ultimate goal of social progress. It was widely believed that "man had the unique ability to guide his fate by controlling the forces of nature through the power of scientific knowledge" (Ravin, 1985).

The Eugenics Committee of the American Breeders Association encouraged the "hard-headed critical and practical study that characterizes the work of our best animal and plant breeders" (Davenport, 1910a). A subcommittee was appointed in 1913 to discover methods for "eliminating defective germplasm from the human population." The report estimated that 10 percent of the population carried germplasm "more or less charged with defects" (Van Wagener, 1914).

The Eugenics Record Office philosophy likewise argued that social progress would occur only with improvements in human germplasm (Rosenberg, 1961). A clinic for the eugenic counseling of prospective parents was planned to guide people in their reproductive lives, but it was never implemented (Allen, 1986). Voluntary efforts at responsible family planning did not appear practicable. Davenport studied the marriage patterns of individuals with hereditary traits such as Huntington chorea and was unable to find any evidence that such people abstained from marriage to block the transmission of the hereditary character (Davenport, 1915b). He urged legislation to prevent such unions: "A state who knows who its choreics are and knows that half of the children of every one of such will become choreic and does not do the obvious thing to prevent the spread of this dire inheritable disease is impotent, stupid and blind, and invites disaster" (Davenport and Muncey, 1916). A. C. Keller argued a similar position at a meeting of the American Academy of Medicine: "Eugenic legislation must turn resolutely to the heavy-handed prohibition of the grosser, more obvious and undesirable phenomena of counter-selection" (Keller, 1910).

Numerous other studies attempted to demonstrate the hereditary nature of many types of mental illness. One report on mental retardation suggested that the controversy over whether this trait did segregate as a Mendelian unit character should not obscure the fact that much mental retardation in the community was hereditary: "This alone would suffice to justify a eugenic campaign . . . The further question of how it is hereditary is purely a technical one" (Feeblemindedness, 1915). Studies of other families appeared to demonstrate that insanity and aggressiveness also behaved in a Mendelian fashion (Rosanoff and Martin, 1915; Key, 1919). Further transmission of such traits should therefore be prevented by state-regulated social isolation or sterilization (Feeblemindedness, 1915). Davenport believed that "inheritable traits are not personal property" and that knowledge about them was important for the good of society. Eugenics, to his mind, was "one of the principal divisions of state sanitation" (Davenport, 1912a).

During this decade professional biologists began to publicly respond to the claims of the eugenics movement, taking the position that knowledge of human genetics was so limited that its application to human reproduction seemed hasty. Kohs felt that the advocacy of a eugenics program based on "slippery and insecure foundations" was a mistake. He feared that the enthusiasts would "let our hopes run riot with our facts" (Kohs, 1915). Conklin also could not support a broad campaign to alter human marriage; the available data on human heredity were so meager that he could only encourage efforts "to eliminate from reproduction the most unfit members of society" (Conklin, 1913).

The overly enthusiastic application of genetic data to human populations also was a problem for W. E. Castle, who recognized the significant changes that such proposals would wreak in the nation's culture:

If society could be managed like a stock farm . . . the average grade of intelligence could be raised by rigid selection long continued . . . But the social consequences of these methods are so tremendous, so subversive are they of individual liberty, that no modern community has ever been willing to contemplate them. Practically we are limited to such eugenic measures as the individual will voluntarily take in the light of the present knowledge of heredity. It will do no good, but only harm, to magnify such knowledge unduly or to conceal its present limitations. (Castle, 1916)

These biologists, then, did not accept the deterministic model for human characters advocated by the eugenics community. Conklin, for example, argued that human characters in germ cells were only potential. Their final outcome always depended upon the interaction between the hereditary units and the environment (Conklin, 1913).

Davenport and his colleagues did examine the relative importance of heredity and environment in many aspects of human disease. In the case of infant mortality, he agreed with the social reformers who believed that improved living conditions could enhance the chances for survival of more babies. But their innate hereditary defects would still be present, Davenport suggested, and these would eventually result in later disease and mortality. In his opinion, "the one fundamentally effective and permanent way of reducing infant mortality . . . is to start the babies in life with good heredity" (Why the babies die, 1918).

The quest for social order and progress was aptly illustrated by ERO testimony before the U.S. Congress after 1920 on the genetic inferiority of certain ethnic groups. The continued influx of large numbers of people from the Orient or southern and eastern Europe was felt by many to endanger the vitality of the existing U.S. stock. The Immigration Restriction Act of 1924 transformed this opinion into law (Furnas, 1969; Ludmerer, 1972; Allen, 1986).

The role of physicians in the eugenics movement at this time was equivocal. D. Fairchild, the president of the American Genetics Association (the successor of the ABA) wrote the editor of the *Journal of the American Medical Association* in 1920 urging cooperation between physicians and geneticists studying human heredity. The goal of further knowledge in this area was a "better and nobler race of human beings" (Fairchild, 1920). Another biologist criticized physicians for paying more attention to their day-to-day work and less to its implication for the future of the race: "Medicine . . . has confined its activities almost entirely to single individuals of its own generation. This science has hardly concerned itself at all with the well-being of future generations. On the contrary, it is bringing to these future generations many evils by its protection of those people who are at present physically or mentally unsound" (Bluhm, 1912). Physicians were urged to act not only as healers but also as judges of persons desiring to reproduce.

During the early part of this decade, however, the opinion that eugenics was an important science was shared by a significant number of U.S. physicians. Franklin argued that the medical profession must become involved in eugenics as a "matter of duty." He believed that parents should make "some effort to see that our descendants not only possess our best, but also that they go beyond us toward the physical ideal" (Franklin, 1913). Hutchinson argued that if the undesirable elements of the population did not reproduce for only two generations, the criminals, prostitutes, mentally ill, epileptics, mentally retarded, and inebriates would be reduced by about three-quarters. "We should look on those people as our unfortunate brothers and sisters who fail to develop, who need our sympathy and protection, not hatred or punishment. Then if we rise to our possibilities, we shall reform society, wipe out crime and reform even our courts and the legal profession" (cited in Davenport, 1912a). Another physician believed that "scientific race breeding in a few generations would empty our jails and insane asylums, banish cancer, tuberculosis and alcoholism, and all physical and mental degeneracy" (Neiberger, 1911).

If, indeed, most human traits—both physical and mental—were controlled by heredity, the presence of defective heredity became an important social issue. In the case of physical characteristics, "If Mendel's law does hold for epilepsy . . . there are important social considerations" (Perry, 1911–12). The same thing was true for families with serious eye diseases such as cataract or blindness, and society should attempt to block the reproduction of such individuals (Howe, 1918, 1919a). The future of the populace was at stake. "Society has the right to protect itself" (Gordon, 1915). On such a hereditarian model, "not only feeblemindedness, alcoholism, insanity, epilepsy, immorality and shiftlessness are hereditary, but so also are superior moral and mental qualities." "The slums do not produce degenerates, but the degenerates produce the slums" (Timme, 1914).

A series of techniques was discussed for improving the genetic health of the community. Students should be educated on the proper selection of marriage partners. Health certificates should be required by the state before marriages could be completed. Segregation in isolated colonies for persons affected with serious diseases should be enforced. Sterilization was recommended in selected cases to inhibit the transmission of serious hereditary diseases (Mack, 1911; Perry, 1911–12; Dixon, 1912–13; Franklin, 1913). For the good of society, physicians should advise against the union of families in which such diseases were evident (Gardner, 1913–14).

To those who found these eugenic efforts toward social improvement harsh and inhumane, the question was put, "Would it be more humane to force into birth human beings who must bear the poverty and mental degradation of imbecility or the many injuries and constant danger which attend epilepsy or the horror of melancholia or delirium, or the moral degradation of habitual drunkenness?" (Wilmanth, 1910–11).

Other physicians were not so confident that social restrictions on reproduction were either appropriate or feasible, given the current state of knowledge in human heredity. Hecht observed that "we know so little of the exact operation of the laws of inheritance with reference to physical characteristics . . . that we have no moral right, it seems to me, to cut down the prospect of the generations that are to come" (Hecht, 1913). The initial fervor of eugenic enthusiasm also decreased as physicians recognized the many impractical aspects of this cause. The identification of individuals with hereditary defects was difficult, and attempts to control their reproduction through state enforcement seemed unworkable. L. Howe at one time advocated colonies to isolate parents of blind children or their voluntary sterilization to prevent the birth of other affected progeny (Howe, 1917, 1919a). He was subsequently appointed chair of the American Medical Association Committee on the Prevention of Hereditary Blindness and collaborated with the ERO staff on an investigation of hereditary eye defects. The committee sought to determine under what circumstances intervention to block reproduction in families with blind children could be justified (Howe, 1919b) and concluded that sterilization of such parents would not be socially acceptable. Segregation would be impracticable. Finally, the committee proposed state legislation that would require persons wishing to marry who had a high likelihood of producing children who were blind to post a bond that would cover the projected cost of their care by society (Howe, 1925, 1926). In this manner it was hoped that individuals and communities would accept responsibility for "their own." H. Laughlin of the ERO expressed this opinion: "When the social unit—family, community, state—which produces a degenerate will be compelled by its neighboring social units to care for that particular degenerate or defective, then there would doubtless develop a sense of responsibility in the producing unit. This, in turn, ought to lead such units to think seriously on measures of prevention" (Laughlin, 1929). Local opinion was thus expected to be more likely to convince such parents not to reproduce than the distant pronouncements of state or federal governments.

The complexity of human beings and the limited data on the precise mechanisms for the heredity of specific human characters were other important reasons that dampened enthusiasm for eugenics among physicians at this time (Waterman, 1920). W. White, an outspoken critic of government-mandated sterilization of affected persons, as early as 1913 stated that "the amount of knowledge of the ancestors of an individual that would make it scientifically justifiable to sterilize him is an amount that is rarely obtainable" (White, 1913). He was not convinced that Mendelian genetics were applicable to much human data, as "unit characters are vague and undefinable and nobody knows what they mean" (White, 1913). Several years later, testifying before the Lunacy Commission of the state of Maryland, he warned that he was "still very much opposed to the whole sterilization movement . . . I do not believe that there is the slightest particle of justification for the mutilating operations that are being advocated broadcast over the country at this time" (White, 1916). And toward the end of the decade, he was still unsure of the general relevance of Mendelian genetics to humans:

> The only theory of the multitudinous theories for why we are what we are, the only one of these theories upon which we may base assumptions that warrant sterilization is the Mendelian theory. Now if the Mendelian theory explained facts with anything remotely resembling the certainty of, for example, the atomic theory or the theory of gravity, or twenty other theories that I might mention, why then we might have something worthwhile to work upon, but when we consider that the Mendelian hypothesis is only one among a number of theories, that there is no unanimity among biologists as to its reliability, and that it has as yet been impossible to apply the hypothesis to man, except to the most limited extent, that even the very

fundamental concepts of the theory, the conceptions of dominance and recessiveness, and the yet more important concept of the unit characters, are vague and indefinable and nobody knows what they mean, it seems to me preposterous that we should endeavor to formulate statutes upon it as a basis . . . The everlasting tinkering with things is what I have no patience with. When we know something different, for God's sake, let us go forward with braveness in that knowledge, but until we know let us have some faith in the powers of nature that have brought us to our present kingdom. (White, 1919)

The modern theory of genetics brought about by the union of cytology and Mendelism was by now familiar to many physicians, who found few human traits to be clearly defined as hereditary unit characters as predicted by the Mendelian laws. Epilepsy, for example, had been viewed by the ERO as a single diagnostic entity that behaved as a simple unit character in different generations of many families as it segregated from one generation to the next. But physicians who dealt with epileptic patients did not accept this overly simplistic explanation, for their experience indicated that many different disorders produced irritation of the brain which resulted in epileptic fits or seizures. This was a symptom, not a specific disease state.

Because most physicians agreed with White that Mendelian unit characters might exist for a few human traits that were probably rare and insignificant for daily clinical practice, they were reluctant to publicly advise a widespread eugenics campaign to prevent reproduction by individuals with so-called defective heredity (Waterman, 1920). The limited information on the hereditary aspects of important human diseases was inadequate to justify wholesale programs to legislate human reproduction.

CONCLUSION

Genetics had become an important part of the biologic sciences in the United States by 1920. Undergraduate and graduate courses in the subject were taught at leading colleges and universities throughout the country, and several medical schools offered courses in human genetics. Research on a wide variety of human characters was undertaken by both biologists and physicians. Most of the work was descriptive, but a few attempts were made by physicians to use animal models in the study of human hereditary disease.

The ERO rapidly became the preeminent human genetic research institution in the United States, and physicians collaborated with its staff to collect pedigrees from families with human disorders ranging from epilepsy and chorea to mental retardation and criminality. A strict hereditarian model for human destiny was explicit in the work produced by the ERO.

Three lines of thought on the role of human heredity in U.S. medicine operated among physicians during this time. The traditional "like begets like" continued to serve as a working definition to explain the familial recurrence of many human characteristics. The direct parent-to-child transmission of a particular trait was accepted as evidence that heredity was an important causative factor.

A second model developed as physicians became aware of the Mendelian theory of heredity and began to consider the possibility that this hereditary mechanism might function in human beings as well as in plants and other animals. There was intense controversy within the medical community as to whether Mendelian heredity could be shown to operate in specific cases of familial disease. Human beings seemed to be so complex in their makeup that whether evidence had in fact been generated to show that this mechanism functioned in human cases remained a serious question.

Because in several instances the pattern of inheritance did appear to match the predictions of the Mendelian theory, a third pattern of thought evolved during this decade, in which both normal and pathologic human characters were interpreted to be dominant, recessive, and sex-linked hereditary traits. Disorders that had previously been labeled "familial" could now be ascribed to the segregation of recessive genetic factors contributed by both the mother and father in the family lineage.

The usefulness of these findings for daily clinical work remained uncertain for many physicians, for these genetic disorders were rarely encountered in clinical practice. They might be interesting from an intellectual viewpoint, but did they have much to do with the wide diversity of human disease which the physician encountered every day?

Questions were also raised about the quality of the information used by research workers to argue that specific human traits segregated in Mendelian patterns. Many of the data were based on incomplete information of family members in different generations. It was obvious that accurate predictions of recurrence risks could not be calculated from such inadequate data.

Human heredity appeared to be more complex than the segregation of simple characters in plants or domestic animals, and premature attempts to label a particular human trait as dominant or recessive without extensive study in many families would discredit human genetics as a quantitative science. Physicians who counseled individuals with such incomplete information might inadvertently injure them as they attempted to guide their reproductive lives.

In the opinion of many physicians, the relative significance of heredity as a cause of important human disease declined during this decade as more illnesses were associated with specific infectious agents. Familial diseases such as cancer or epilepsy, previously felt to have strong hereditary predispositions, now appeared to result from infection by specific bacteria. Heredity might be responsible for a few rare human diseases, but the available scientific evidence argued instead that the suffering of millions of people was the result of infectious agents. Hereditary disease had traditionally been viewed by physicians as inevitable and therefore untreatable. Now there was new hope that physicians could make a difference in the conquest of disease by diagnosing and effectively treating the myriad of infections that afflicted the human race.

In contrast, numerous reports from the ERO argued that many human ills had a strong hereditary component. These research workers in human genetics feared that uncontrolled breeding of the undesirable elements would have grave implications for the future of U.S. society, a concern shared by several prominent physicians. For the good of society, individuals with hereditary defects were advised not to marry. If such voluntary efforts failed, then state-enforced segregation or sterilization had to be considered, as society sought to protect itself.

The majority of U.S. physicians recognized the preliminary nature of the available data on the heredity of human traits. Pedigrees had been collected on only a limited number of families. Different patterns of inheritance occurred in different families segregating the same characteristic. The accuracy of the Mendelian laws for predicting recurrence risks was simply not known.

By the end of the decade, enthusiasm for social interference with human reproduction was at a low ebb among most physicians. The importance of heredity as a cause for human disease was in question. The ability of the leading theory of heredity to predict the outcomes of matings in specific families was unclear. There were too many unanswered questions in this area to justify the wholesale attempt to regulate human reproduction.

SIX

ON THE MAKING OF PARADIGMS

Our views on heredity have been profoundly modified by the studies of Weismann, Mendel and others. —Osler, 1907

The major chronic diseases of our time including cancer all have impor- tant genetic components, and cancer itself is essentially a genetic disease at the cellular level. The Human Genome Project, whose aim is to map the position of all the functional genes and eventually to sequence the whole human genome, will provide the basis for dealing with these diseases in the next century. —Bodmer, 1991b

While U.S. physicians of the late nineteenth and early twentieth centu- ries have often been portrayed as poorly educated artisans, tradespeo- ple of medical practice, my review of the historical data suggests just the opposite: that many physicians were in fact quite knowledgeable about changes within biology and particularly within the new science of genetics as it evolved after 1900. They listened to lectures on these new developments and discussed their possible implications at local and national medical society meetings, and they read the many articles on them in the medical literature. They wondered out loud whether these scientific findings were applicable to their clinical practices. Would this revolution in scientific thinking be useful to them as they cared for their patients?

THE EVOLUTION OF SCIENTIFIC THEORIES

The day-to-day work of modern scientists is driven by their intense personal curiosity to learn more about the world around them. The scientific method is not a random assemblage of facts about the natural world but instead a human-guided enterprise to both collect data and interpret them. Facts in science are said to be theory laden. Their meaning depends upon the mind-set of the scientist who goes to the trouble to collect them in the first place (Hukk, 1988; Bauer, 1992).

Scientists may start their quest for discovery either from experiment or from a concept. An observed phenomenon may suggest a certain hypothetical explanation, or a hypothesis may suggest specific planned observations or experiments whose results may have been predicted by the hypothesis (Humphreys, 1968). Three elements have been broadly identified with the daily work of modern scientists. First, personal curiosity drives them to seek an understanding of how the natural world functions; through observations, a theoretic explanation of a phenomenon is postulated. Second, this explanation is communicated to colleagues in science. Finally, other scientists assess the usefulness of this contribution by either confirming or denying its ability to predict the outcomes of future experiments (Hukk, 1988).

Many philosophers of science have argued that there can be no logical explanation for discovery. But once a new idea becomes available, philosophical constructs can be utilized to analyze its acceptance or rejection by the general body of scientific knowledge (Darden, 1991). New scientific theories rarely spring fully formed into the minds of individual scientists. Instead, they evolve in stuttering steps toward a better-defined explanation of certain natural phenomena. New ideas, often vague, are tested experimentally to determine whether they can predict a particular outcome. Negative results might suggest that the entire theory should be rejected, but in the real world of science, they often instead suggest refinements that improve the theory under consideration. The theory's ability to resolve anomalous outcomes is a test of its general adequacy to explain a wide range of natural phenomena (Darden, 1991).

Once a new theory has been shown to be useful by different scientists in a particular field of endeavor, it becomes part of the accepted basis for explaining how the world works. Much of the daily work of scientists involves using accepted theories to make predictions and then attempting to verify the expected outcomes by experimentation. Thus, scientists seek new knowledge within the confines of known theoretical foundations, a process often called "normal science." Working scientists do not seek novel explanations for natural phenomena, for these are initially untidy and often appear illogical and hence "unscientific" (Bauer, 1992).

Theories in science, then, function as guides to reality as perceived by the scientists. Suggested changes in theory are usually resisted by scientists because they imply that previously accepted principles were inadequate (Bauer, 1992). A conservative system of beliefs about how the natural world functions, science would be threatened by the easy acceptance of radical new theories of scientific explanation. If experimental data do not agree with accepted theory, most scientists choose to perform more experiments that might confirm the theory rather than to seriously question its validity. Scientists seek consistency in their work, which keeps them on track for routine, day-to-day work (Bauer, 1992).

But the accepted explanations occasionally fail. A scientific anomaly is defined as an unexpected observation that is difficult to explain within the context of existing theory. Scientists may respond to anomalies in several different ways. The anomaly may simply be ignored, assuming that at some later time a new theory may be proposed based upon other experimental work which in retrospect will provide a logical explanation for the neglected anomaly. Or the anomaly may encourage the formulation of a new theory, if the existing ones cannot be modified to encompass it (Darden, 1991; Lightman and Gingerich, 1991).

Science is a human enterprise, and scientists react to change as other human beings do. They are uncomfortable with change and often resist new theories, even when these may provide better explanations for natural phenomena. New knowledge from a frontier area of science is controversial, fragile, inadequately tested, and perhaps altogether wrong (Bauer, 1992). Scientists respond to this questioning of their belief system in predictable fashion: "The existence of dissonance, being psychologically uncomfortable, will motivate the person to try to reduce the dissonance and achieve consonance. When dissonance is present, in addition to reducing it, the person will actively avoid situations and information which would likely increase the dissonance" (Festinger, 1957). Scientific innovation may thus involve uncomfortable periods in which existing theories no longer seem to work and new theories are being proposed and refined. Only after a period of conflict and eventual resolution as a new theory gains acceptance can the anomalies be explained and "safe" daily science resumed by the working practitioners.

The paradigm model for the metamorphosis of scientific concepts developed by Kuhn has had a profound effect on the historical study of science over the past twenty years. Kuhn identified two distinct phases of scientific work. First, *normal science*, which involved the extension of knowledge predicted by the current paradigm or leading hypothesis of the day. Facts were gathered by practicing scientists to confirm predictions of the paradigm model. Occasionally data appeared which were at odds with these predictions. Attempts were then made to explain these anomalies using the paradigm. If this could not be done successfully, a crisis in scientific thought developed. In the second phase, the creative process or *revolutionary science* then sought to develop a better model to explain the new observations. Several competing new paradigms were often proposed, but none would be rapidly accepted by the scientific community. Practitioners had used the traditional model for years and would accept a new one only after a period of conceptual rethinking, eventually adopting the new paradigm both for its improved problem-solving ability and for its simplicity or "neatness." Normal science could then resume its daily task of filling in the knowledge gaps created by the new paradigm (Kuhn, 1970).

Kuhn's analysis of revolutions in scientific thought used examples drawn mostly from the physical sciences. Although in his scheme, a new paradigm almost always replaced the previous one, a recent application of this model to the development of human genetics since 1930 argued that this science has evolved more by a process of paradigm fusion than by replacement. Several lines of thought from biochemistry, cytology, and cytogenetics often came together to produce a new paradigm. None of the individual paradigms was replaced, but the hybrid paradigm was more robust than any one alone and could explain more of the questions raised by the genetic community of the day. Modern human gene mapping, for example, has developed as a combination of developments within human cytogenetics, somatic cell hybridization, and the recombinant DNA technology of restriction fragment length polymorphisms (Dronamraju, 1989).

But radical change in science never comes easily. Darden outlined various schemes used by working scientists to modify existing theories when confronted by an anomaly. Paradigm shifts occur only when scientists are unable to resolve the anomaly in the context of existing theory (Darden, 1991). Intratheoretic explanations are always given precedence over intertheoretic ones, which require a wholly new thought construct. Scientific innovation becomes acceptable only when no satisfactory explanation for the observed natural phenomenon can be drawn from the existing theoretic framework (Humphreys, 1968).

U.S. physicians in the early twentieth century were confronted with competing theories to explain the causes of human diseases, which they reviewed in the light of expected results and their personal clinical observations, finally choosing the theoretic construct that seemed to work best for them.

THE DECLINE OF THE GENETIC PARADIGM IN U.S. MEDICINE

Conceptual shifts are evident in the analysis of human genetics within the context of U.S. medicine during the period 1860 to 1920. The normal science phase involved the traditional definition of heredity used before 1900. Direct transmission of a character from parent to child implied that heredity was at work: like begets like. The absence of such direct transmission indicated that some other mechanism was necessary to explain the symptoms of familial diseases. Most practitioners agreed that specific disease itself was rarely hereditary. Instead, a predisposition or diathesis to disease was the entity actually passed from one generation to the next. Symptoms of disease developed after an external trigger acted on vulnerable tissues to produce the abnormal functioning diagnosed as disease. With this understanding of diathesis, affected individuals within a family could have encountered different triggers over time and hence might have developed different symptoms. For example, epilepsy in a parent might result in insanity or alcoholism in a child.

The crisis phase of the diathesis model had actually begun before the rediscovery of Mendel's work in 1900. Physicians had become better educated in science in general and had begun to recognize that the traditional paradigm for heredity explained much but was untestable from an experimental point of view. It predicted everything, yet nothing specific. Diatheses ran in certain families, but how did that actually result in disease, and did it have any predictive value for physicians as they tried to counsel prospective parents from such families?

The rapid developments within genetics during the first decade of the new century were well known to many physicians in the United States. Did the new science, which appeared to explain much of animal and plant inheritance, have any relevance to normal and pathological human characters? Mendel's laws were mentioned in many reports of human disease at this time as possibly germane, but no one in medicine was sure how these statistical constructs could explain the familial aggregation of human diseases. During this period of transition, physicians used the traditional understanding of human heredity at the same time that they pondered the importance of the new models.

An example of this was Loeb's use of Ribot's 1875 genetic model for his analysis of familial blindness in 1909. In historical retrospective, it is easy to fault him for employing such an antiquated model for heredity. Because his article, like many medical articles of the day, had a limited bibliography, it is impossible to determine whether he considered the Mendelian model and then rejected it, or whether he was simply unaware of it and used the model for heredity which had worked in the past. In any event, his colleagues in ophthalmology did no better. None of them was able to apply the Mendelian model to the inheritance of eye diseases until several years later.

The Mendelian paradigm for human genetics was eventually understood by many physicians after 1910, and it was then used to explain the segregation of a number of familial human disorders. Most of these were rare traits, but the applicability of the theory appeared to work in such diverse specialties as hematology, ophthalmology, orthopedics, and urology. Some physicians believed that it worked so well that there was a danger of over-Mendelization. It was possible to overreach the available data and assume that virtually all human traits were inherited as simple Mendelian unit characters.

A more balanced view of the importance of heredity in humans required a thoughtful application of the theory to the actual pedigrees in which varied traits segregated. Collaborative work between physicians and scientists was performed on a limited basis between 1910 and 1920, particularly in ophthalmology and neurology. But at the same time that physicians' interest in human genetics increased, the attention of many scientists was diverted to other directions. The study of basic genetic mechanisms focused primarily on experimental breedings of plants and animals. Most scientists applied their efforts to the type of research work that appeared to have the greatest likelihood of success. At this time the leading areas of research involved plants and the fruit fly Drosophila (Ludmerer, 1972; Glass, 1986).

Human beings had proven to be difficult experimental subjects. Genetic techniques that worked well in plants and animals produced only muddled data when applied to human pedigrees. Such work on humans was called "clumsy and cumbersome" (Dronamraju, 1989). Much human genetic work after 1910 involved pedigree collection and the explanation of segregation patterns based on analogous situations in other mammalian species (Ludmerer, 1972; Glass, 1986).

By the 1920s the importance of human genetics for the leading scientists was waning. T. H. Morgan in the Mellon Lecture at the University of Pittsburgh School of Medicine in 1924 presented a review of the current understanding of Mendelian genetics as demonstrated by detailed study of plants and animals. He mentioned several rare human traits that appeared to segregate in a Mendelian fashion as well: brachydactyly was dominant, albinism was recessive, and hemophilia and color blindness were sex-linked. Despite the fact that he was speaking before an audience of medical students and physicians Morgan did not discuss the possible application of the new genetics to human disease processes. Nor did he encourage his audience to study the segregation of human traits that appeared frequently in certain families (Morgan, 1924).

At the same time that the professional geneticists demonstrated decreasing interest in human genetics, several physicians developed a lively curiosity in human heredity and argued that the Mendelian principles did apply to many human disorders. Herrman discussed heredity in clinical pediatrics before the New York Physicians Association in 1923, presenting examples of normal human traits that appeared to be controlled by Mendelian factors segregating within families. He believed that hair form, eye color, physiognomy, and intelligence all were inherited. Disease characters also behaved like Mendelian factors: abnormal bone growths, diabetes, cleft lip and palate, and nervous and ocular disorders all appeared to have a significant genetic component that caused symptoms of disease (Herrman, 1924).

There were several reasons for the difference in emphasis regarding human heredity in the science of genetics and the practice of medicine. Geneticists and physicians rarely communicated, and neither evinced much appreciation of how the other worked or what the other's professional goals were. Scientists often criticized physicians for looking at one case at a time and not developing a societywide perspective on human heredity: "Medicine . . . has confined its activities almost entirely to single individuals of its own generation. This science has hardly concerned itself at all with the well-being of future generations. On the contrary, it is bringing to the future generations many evils by its protection of those people who are at present physically or mentally unsound" (Bluhm, 1912). The calling of physicians as healers and comforters of the sick was therefore felt to be less important than their newly assigned task as judges of individuals desiring to reproduce.

Physicians recognized society's wish to prevent evil, whether physical, moral, or mental, but they continued to view medicine as a personal interaction between physician and patient, not as one aspect of a public health program: "In the physician's zeal to do something great—to take part in some big movement to prevent disease wholesale—we must not lose sight of the patient as an individual" (Wright, 1914). The physician also sought to engender hope in the minds of sick people. A hereditarian paradigm in which much disease was inevitable destroyed this comforting aspect of medical practice. Genetics then became a "dismal" science (Inheritance of acquired characters, 1919; Snyder, 1951).

After 1915 the scientists doing research on human genetics were for the most part eugenicists who sought to apply genetics to the improvement of society at large. But there was increasing dissatisfaction in both the scientific and medical communities with the political aspirations of these workers and with the quality of their scientific output. Many scientists and physicians viewed those within the eugenics campaign as something of a "lunatic-fringe" that would ultimately have little importance (Glass, 1986). Still, the ability of the ERO staff to convince congressional leaders of the necessity to protect the United States from foreign germplasm was unmistakable. The passage of the 1924 Immigration Restriction Act forced other biologists to reconsider the claims made by the eugenics community (Ludmerer, 1972).

Davenport and his colleagues produced some solid research on human genetics after 1920. Their review of sex-linked human traits was a careful survey of eleven characters that followed this pattern of inheritance, and they also studied the linkage between two traits (hemophilia and color blindness) in individuals of a family through five generations (Davenport, 1930).

New ideas never exist in a vacuum. They flourish or die as their cause is championed by eminent individuals within the scientific community. The eugenic aspects of human genetics thus seemed to be correct, because Davenport spoke and wrote forcefully on this topic before the general public as well as the medical and scientific communities. The power of his personality is demonstrated by his attracting almost a million dollars in donated money to support the work of the ERO between 1910 and 1930 (Allen, 1986).

No comparable spokesperson for the importance of genetics arose in U.S. medicine. None of the leading physicians of the era was convinced that these new ideas on human heredity would have important consequences for modern medical practice. "Scientific ideas compete in an open marketplace. Each offers the possibility of a plausible solution to what might be a potentially significant problem. In its promise, an idea will attract other scientists—fellow explorers who will articulate, criticize and ultimately determine the idea's actuality. While these explorers can breathe life into an idea, their absence or defection leads to its death" (Krantz and Wiggins, 1973). In summation, "the greatest danger for any new idea is for it to be ignored" (Hukk, 1988). This was to be the eventual fate of the new genetics as applied to clinical work in U.S. medicine.

Criticism of the ERO program after 1920 was broad based; it came from both the medical and the scientific communities. Davenport had urged physicians to submit interesting pedigrees with diverse human traits to his staff for analysis, but he became irritated by comments on the quality of the data collection performed by his own field-workers. Such criticism, he complained, came from "physicians and is part of an assumption of superiority that is so widespread among medical men and is so particularly emphasized by some of them, that one is almost led to suspect that it is the result of an understanding in the profession" (Finlayson, 1916).

Biologists such as Morgan, Jennings, Pearl, Castle, and Muller also began to pay more attention to human genetics, particularly as it appeared to have a significant public impact. They argued that the claims of the eugenicists that most human traits were controlled by single Mendelian unit characters were simply unsubstantiated by the available data on human inheritance. The data collection system had flaws, and the acceptance of such data in an uncritical manner was not good science (Little, 1922; Allen, 1986).

This research work seemed to involve a great deal of circular reasoning. The ERO collected data from families with "genetic" disorders to confirm that the trait in question was indeed hereditary. Collection of data from other families then confirmed these preconceived expectations. No contradictory data were available because there were virtually no other scientists working in the field of human genetics. From the standpoint of Davenport and his co-workers, there was internal consistency in their results. Their theory that many human characteristics—physical, mental, and moral—were controlled by genetic elements was amply supported by their observations from the family data. Their evidence remained uncontradicted and thus continued to support their theory. Additional study confirmed the theory, as the expected outcome was repeatedly discovered. But successful prediction does not necessarily prove that the theory itself is correct (Bauer, 1992).

Internal reviews of the ERO by its parent organization, the Carnegie Institute of Washington, in 1929 and 1935 also found serious problems with the quality of the pedigree data that had been collected over the years. The records were often incomplete and judged to be "unsatisfactory for the study of human genetics" (Allen, 1986). Funding for the ERO was gradually withdrawn, and the institution closed in 1939. The work of these leaders in the field of human genetics had progressed to the point where it seemed "hardly to be science at all" (Haller, 1984).

Physicians had always looked to the ERO as the major source of information on human genetics in the country. The repudiation of its work by the scientific community implied to many physicians of the day that human genetics was not particularly important.

From the viewpoint of the physician after 1920, genetics appeared to have little impact on the important diseases. The medical successes of the age were strictly environmental. Etiologies of specific diseases appeared to be exogenous, and it was no longer necessary to postulate a diathesis or predisposition to disease (Ackerknecht, 1982). Improved sanitation and vaccination prevented many serious diseases, while the use of antisera halted the progression of other previously lethal infections. Physicians at the time were said to display a "marked lack of enthusiasm for genetics" (Hobshawn, 1987). William Osler, for example, who had expressed a great interest in genetics early in the century, in the second edition of *Modern Medicine* in 1915 deleted any discussion of the relationship between heredity and human disease. L. Barker from Johns Hopkins commented on this shift in attention: "There surely must be special reasons for the glaring contrast between the fiery enthusiasm for research in heredity and development evident in the biological laboratories of every university and coolness toward genetic questions and the concentration upon environmental influences that characterize activities in hospitals and even in the laboratories of clinical research in our university medical schools" (Barker, 1927). Most physicians at this time agreed that heredity might play a role in some rare human conditions, which Ludmerer called "academic playthings of no social or practical importance" (Ludmerer, 1972). Each character was so uncommon that the average practitioner might never encounter one such case during an entire professional career.

On the other hand, any practicing physician could recall examples of familial disease. Macklin reported a discussion with twenty-five physicians on the role of heredity and disease in which they remembered familial examples of heart block, renal stones, anemia, dislocated lenses, and mental deficiency encountered over the years. None of these data had been reported in the medical literature, because not one of the physicians believed that they were important (Macklin, 1933).

This skeptical nature on the part of the practitioner would persist for many years. One commentator in 1940 noted, "It seems surprising that so few medical men are aware of the possibilities which the increasing knowledge in this field [of human genetics] holds for them . . . Even today medical genetics, as a special subject, has not yet found its way into the curriculum of more than two or three medical colleges in this country. While the United States has led the world in experimental genetics, we have allowed the initiative in medical genetics to be taken by a number of other countries" (Scheinfeld, 1940).

CONTEMPORARY GENETICS AND AN INTERACTIONAL MODEL FOR THE CAUSES OF DISEASES

My historical analysis of the evolution of ideas on genetics in U.S. medicine to 1920 has documented the shift in opinion from the early assumption that much human disease had a significant hereditary component to the more modern notion that this was rarely the case and that most important disease states were the result of environmental factors. The interaction between the state of the organism and external triggers had been recognized for many years before this time, but most physicians had largely discarded the notion of a diathesis or predisposition to disease by 1920 (Herrman, 1918; Ackerknecht, 1982).* In subsequent years, this environmental paradigm for disease causality enjoyed enormous success in the explanation and eventual control of many health menaces from tuberculosis, cholera, and typhoid to atherosclerotic heart disease.

Simultaneous developments in human genetics since 1950 have elucidated the biochemical defects in hundreds of rare single-gene disorders such as phenylketonuria, sickle cell anemia, and Tay-Sachs disease, the "academic playthings" noted earlier. Altered gene products such as enzymes or hemoglobins have been isolated and characterized. Mutations within the relevant genes which coded for these materials have then been postulated based upon the altered amino acid sequence of the gene products.

But since 1980, new recombinant DNA techniques have permitted the development of "reverse genetics." In this approach the abnormal gene itself is first isolated and its nucleotide sequence determined. The gene product is then sought from affected tissues, and its altered bio-

---

*Breeding studies in experimental animals had shown by 1920 that susceptibility to particular bacterial infections segregated as a Mendelian character (Hagedoorn and Hagedoorn, 1920). These observations appear to have had no impact on the thinking of clinicians at that time.

chemistry within the cell studied to understand how dysfunction resulted in disease in affected individuals. Two recent examples illustrate the spectacular success of this paradigm. The mutant gene in Duchenne pseudohypertrophic muscular dystrophy has been isolated from the X chromosome. Its gene product, a muscle protein called dystrophin, has now been isolated from both healthy and diseased tissue (Koenig, Hoffman, and Bertelson, 1987; Hoffman, Fischbeck, and Brown, 1988). Similarly the cystic fibrosis gene has been isolated from chromosome 7. Its gene product, the cystic fibrosis transmembrane conductance regulator, has also been characterized (Riordan, Rommens, and Kerem, 1989; Rommens, Iannuzzi, and Kerem, 1989). The detailed biochemistry of both these gene products is an area of intense investigation into how the altered functioning of these proteins produces symptoms of clinical disease. It is hoped that this knowledge will provide further clues for the effective treatment of these lethal inherited diseases.

These rapid changes in biology may now signal a change in the way research is done. One model suggests that a scientist will soon look at DNA nucleotide sequences on a computer screen (nucleotides are the basic structural units of DNA), choose a region that looks interesting, order the gene product from a reference laboratory, and then study what it does *in vitro* (Gilbert, 1991). Such an approach to research has become plausible because of the data now being generated by the Human Genome Project, a multinational fifteen-year undertaking to provide a detailed genetic map of the entire human genome, comprising about three billion nucleotides. The research strategy involves several levels of genome mapping. First, polymorphic markers are to be identified at an average distance initially of 10 cM (centimorgans), with subsequent refinements expected to provide markers at an average distance of 2 cM. Second, overlapping clones of DNA between two and eight million nucleotides are to be prepared and the sequence of each determined. Third, the project ultimately means to determine the complete nucleotide sequence of the human genome, but available sequencing techniques are too slow and too costly to achieve this target within the fifteen years projected, so the development of new sequencing technologies will be encouraged. The application of such technics to model organisms with smaller genomes is expected to enhance the human genome project. Fourth, sequencing efforts are already under way to study the genome of several model organisms. The bacterium *Escherichia coli* with five million nucleotides, and the primitive eukaryote yeast *Saccharomyces cerevisae* with fifteen million nucleotides, are the subjects of intense investigation. Multicellular organisms will be examined next. The nematode *Caenorhabditis elegans* with 100 million nucleotides and the fruit fly *Drosophila melanogaster* with 160 million nucleotides provide organisms of intermediate genome size. If efficient technics for the rapid sequencing of DNA can be perfected on these model systems, it is hoped that the human genome will not prove to be an insurmountable obstacle. Sequence homologies between the various species will also aid in the interpretation of the function of genetic sequences as these data from the human material become available (Bishop and Waldholz, 1990; Watson, 1990; Green and Waterston, 1991; Jordan, 1992).

Within the field of human genetics, it is today generally agreed that the application of this knowledge will engender the development of new paradigms on the mechanisms of inheritance and gene expression (Dronamraju, 1989). I believe that the availability of such detailed human genetic information will produce an important paradigm shift within clinical medicine as well, as a recent comment from the Washington University Medical Center also suggests: "During the next generation a fundamental transformation in medical science will affect all our lives. Think of everything blood transfusions, antisepsis and anaesthetics did for surgery. This will be bigger. Recall what vaccines and antibiotics did to control cholera, typhoid and other horrible infectious diseases. This will be more important" (Sutter, 1991).

Family studies have indicated for many years that common disorders such as cancer, heart disease, mental illness, and diabetes appeared to have some hereditary component. The availability of genomic data will encourage new efforts to correlate changes within the sequence of DNA with the development of particular human diseases (Marx, 1990; Watson and Cook-Deegan, 1990). A reassessment of the nature-versus-nurture controversy is becoming inevitable.

The new paradigm will pay more attention to the complex interaction between intrinsic genetic and extrinsic environmental factors that together produce the altered functioning labeled "disease." The nineteenth-century notion of diathesis or predisposition to disease is having a revival, as a latent paradigm comes to the fore once again. It is now called genetic susceptibility to disease. A 1992 MEDLINE computer bibliographic search of the National Library of Medicine database of articles in English from the past eighteen months yielded more than four hundred relevant articles on a range of common physical and mental disorders whose etiology appears to have a genetic basis to some degree (table 6.1). The detailed picture of the human genetic structure

Table 6.1.   Selected Studies in the Genetic Susceptibility to Disease, 1989–1992

Disorder	Reference
Alcoholism	Blum, Noble, and Sheridan, 1990
	Holden, 1991
Alzheimer disease	St. George-Hyslop, 1990
Arthritides	Nickerson, Luthra, and David, 1990
Autoimmune diseases	French and Dawkins, 1990
Cancer	
Breast	Hall, Lee, and Newman, 1990
	Malkin, Li, and Strong, 1990
	Claus, Risch, and Thompson, 1991
	M. C. King, 1991
	Hall, Friedman, and Guenther, 1992
	Margaritte, Bonati-Pellie, and King, 1992
Colorectal	Shike, Winower, and Greenwald, 1990
	Kinzler, Nilbert, and Vogelstein, 1991
Lung	Petersen, McKinney, and Ikeya, 1991
Melanoma	Kleeburg, 1989
	Yoshikawa, Rae, and Bruins-Slot, 1990
Retinoblastoma	McKee, Yandell, and Dryja, 1990
	Weinberg, 1991
Wilm tumor	Huber, Buckler, and Glaser, 1990
Coeliac disease	Kagenoff, 1990
Diabetes mellitus	Dawkins, Martin, and Saueracker, 1990
	Erlich, Bugawan, and Scharf, 1990
	Jenkins, Mijuvic, and Fletcher, 1990
	Nepom, 1990
	Tait, 1990
	Hyer, Juler, and Buckley, 1991
Drug toxicity	Ruchelli, Horn, and Taylor, 1990
	Weber, 1990
Hashimoto thyroiditis	Badenhoop, 1990
HIV infection AIDS	Fabio, Smeraldi, and Girgeri, 1990
Hypertension	Williams, Hunt, and Husstedt, 1989
Lupus erythematosis	Franek, Timmerman, and Alper, 1990
Multiple sclerosis	Oksenberg and Steinman, 1989–90
	Lord, O'Farrell, and Staunton, 1990
Tuberculosis	Khomenko, Litvirov, and Chukamova, 1990

which will emerge over the next decade will undoubtedly encourage further attempts to investigate changes within the DNA which alter normal cellular functioning and may eventually produce symptoms of specific diseases.

A paradigm that incorporates the interaction between genetic and environmental factors will emerge as the predominant model for mechanisms of human disease in the twenty-first century. There will certainly be a spectrum of relative importance for each of these two competing forces:

1. In rare instances an alteration in the nucleotide sequence of a particular gene may result in an abnormal gene product that is critical for the normal functioning of a particular type of cell. In such a case, clinical disease becomes inevitable, regardless of environmental factors. The gene product in Huntington chorea, for example, remains poorly characterized, but the presence of the abnormal gene invariably produces signs of neurologic degeneration in affected individuals if they live long enough. Likewise, the mutation within the beta-globin gene which codes for sickle hemoglobin always results in abnormal functioning of the red blood cells and clinical symptoms of severe anemia.

2. At the opposite pole of the nature-and-nurture controversy are examples of diseases brought about by environmental factors whose influence cannot be ameliorated by the function of normal human genes. Clinical disease then always develops, regardless of the genetic endowment of the affected individual. Rabies infection, certain snake venoms, and high-dose X rays all damage specific tissues to such an extent that the physical integrity of the target individual is seriously compromised and death almost always occurs.

3. In the larger sphere of most human disease, specific people have an altered genetic structure that results in clinical disease under specific environmental influences. The absence of the initiating trigger allows normal body functioning and hence the absence of discernible disease. For example, in persons with phenylketonuria a mutation in the gene for phenylalanine hydroxylase blocks the metabolism of the amino acid phenylalanine. When the individual at risk consumes "normal" amounts of phenylalanine, the concentration of this material increases in body tissues, eventually damaging the central nervous system and producing seizures and mental retardation.

Individuals with specific alterations in genes controlling immunologic function which make them more susceptible to infection by bacteria and viruses have been identified in familial clusters of AIDS and tuberculosis (Fabio, Smeraldi, and Girgeri, 1990; Khomenko, Litvirov, and Chukamova, 1990).

On a nucleotide level, the current model for the development of the malignant eye tumors in retinoblastoma illustrates the interaction between genetics and environment. The RB gene is located on chromosome 13 and codes for a protein, pRB, which appears to regulate transcription of messenger RNA. In familial retinoblastoma, both copies of the RB gene have mutations that block expression in the cells of the eye. This loss of transcriptional control alters normal cellular growth processes and permits the unregulated cell division that is characteristic of malignant tumors. In nonfamilial cases, acquired mutations in both genes again prevent the expression of this important regulatory protein, and abnormal cell growth and tumor formation ensue (Weinberg, 1991).

Research on other malignant tissues has demonstrated several features of tumorigenesis which result from an interaction of internal and external triggers. Many tumors appear to be initiated by genetic changes within somatic cells. Some of these involve the rearrangement of chromosomes via deletion or translocation of specific nucleotide sequences. Clinical cancer eventually results from dedifferentiation, the loss of normal growth control of the involved tissue (Bodmer, 1991a). In human neoplasms more than two hundred chromosome anomalies have been associated with the development of specific malignancies. Other chromosomal abnormalities have been observed which develop after the tumors have begun to form. These may be essential for the unregulated malignant growth of the particular tumor (Mitelman, 1991).

A model for the initiation and evolution of colon cancer has been proposed which exemplifies many features of the interactional model for human disease. Endoscopic surveillance of the colon in patients who eventually develop overt carcinoma has shown a fairly predictable sequence of events which alters normal mucosal cells. First, various types of benign tumors or polyps develop. Malignant degeneration of these then results in frank carcinoma and metastatic spread to other abdominal organs. This multiple-step process correlates with successive alterations in the genome of the colon mucosal cells (Marx, 1989; Weinberg, 1991).

It has long been postulated that heredity played an important role in the development of certain types of bowel cancer (Wartin, 1913). Familial adenomatous polyposis (FAP) is an autosomal dominant disorder in which affected individuals develop hundreds of polyps within the colon. It has served as a model for understanding the evolution of bowel cancer in general. Malignant transformation of the polyps usually begins between age twenty and thirty. The gene involved in FAP has been mapped to chromosome 5 and has been labeled the APC locus. In nonfamilial cases of colon cancer, the loss of the APC gene function also appears to be involved in the earliest stages of carcinogenesis. A related gene, MCC, located nearby on chromosome 5, is mutated in both familial and sporadic colon cancer. The loss of this gene function permits increased cell division (Lindgren, Bryke, and Ozcelik, 1992).

Hereditary nonpolyposis colorectal carcinoma (HNPCC) is a second genetic disorder which accounts for between 4 and 13 percent of all bowel cancers. Affected individuals often develop malignancy in other organs as well, especially the stomach, uterus, biliary tract, pancreas, and urinary tract. A genetic locus on chromosome 2 has been associated with HNPCC. The gene is not deleted in malignant cells, and it appears to induce mutations in many other loci on different chromosomes. The gene product may affect the accuracy of DNA replication and hence produce widespread genomic instability. Normal cell physiology may be altered to the point where unregulated growth, or malignancy, eventually results in many different tissues. Instability at multiple genetic loci has also been associated with other hereditary disorders that yield malignancies, such as Bloom syndrome, xeroderma pigmentosum, and ataxia-telangiectasia (Aaltonen, Peltomaki, and Leach, 1993; Marx, 1993; Peltomaki, Aaltonen, and Sistonen, 1993; Thibodeau, Bren, and Schmid, 1993).

Another early change in the evolution of colon tumors involves increased activity of the enzyme methyltransferase. This gene product removes methyl groups from cellular DNA. Such a change allows for the enhanced expression of DNA and may permit dedifferentiation of normal mucosal cells (El-Deiry, Nelkin, and Celano, 1991). The k-ras oncogene (a mutant of the gene that regulates cell growth rate) is then activated on chromosome 12, again permitting unregulated cellular growth (Bodmer, 1991a). The function of the DCC gene on chromosome 18 is another important factor in malignant degeneration in the colon. Different alterations within this gene, such as deletions or insertions, have been found in different tumor cell lines. The amino acid sequence of the DCC gene product suggests that it may function as a cell adhesion surface glycoprotein. Loss of this gene function may alter cell-cell adhesion in the colon mucosa, again permitting dedifferentiation of the normal tissue (Fearon, Cho, and Nigro, 1990; Weinberg, 1991). And a late stage in the evolution from benign to malignant tumors in the colon involves the loss of the protein from the p53 gene on chromosome 17. The evolution of normal tissue to frank carcinoma therefore involves multiple genetic changes within somatic cell lines. Both the inactivation of suppressor genes and the activation of oncogenes are involved as tumors evolve toward malignancy (Bodmer, 1991a; Weinberg, 1991).

The model proposed for the development of colon carcinoma is not strictly reductionistic or inevitable. Both clinical and laboratory observations suggest factors that can modify the progression of this disease. In families segregating FAP, individuals who inherit the same mutant gene may have varying degrees of malignant degeneration at the same age. In one family, monozygotic twins with FAP had developed varying numbers of bowel polyps. One twin had six tumors at the same time the other had thirty-seven. Other genetic modifiers or environmental factors have been suggested to explain the marked variability of disease in such individuals (Spiro, Otterud, and Stauffer, 1992). In fact, both types of influence probably are involved. A mouse model for FAP has been developed in which unlinked modifier genes clearly affect the number of bowel tumors in animals carrying the APC gene (Su, Kinzler, and Vogelstein, 1992). Clinical observations have documented the importance of environmental factors in the development of tumors in susceptible individuals. The ingestion of increased dietary fiber has slowed the evolution of malignant polyps in people carrying the FAP gene (DeCosse, Miller, and Lesser, 1989).

Different levels of explanation for the development of colon carcinoma were elucidated in a recent study by Potter, who attempted to correlate genetic and environmental factors shown to influence the process. Population studies have linked high levels of dietary fat intake with colon cancer. High vegetable fiber intake, on the other hand, lowered the risk. Persons with elevated levels of bile acids also had increased cancer risk. Increased fat in the diet enhances the secretion of bile acids from the liver and thus increases the exposure of the bowel mucosa to these potentially toxic materials. Dietary fiber binds bile acids and may inhibit their toxic effects on the mucosa. Vegetables also contain a number of nutrients that induce enzymes in the mucosa to inactivate other potential carcinogens from the diet.

Other dietary materials may alter the activity of DNA methylation in the mucosa. This may then inhibit one of the earliest stages involved in malignant transformation. The cell adhesion molecules, such as the one coded by the DCC gene, all depend on calcium for their normal action. Decreased dietary calcium could potentially enhance cell disruption and facilitate the loss of differentiation within the bowel lining. Activation of oncogenes may also be influenced by dietary factors. Diacylglycerol, a component of dietary fat, appears to function in the transmembrane signaling system, which eventually activates oncogenes such as ras. Elevated levels of fat in the diet may therefore accelerate this process toward malignant degeneration (Potter, 1992).

This type of model for human disease, which integrates environmental and genetic elements, will become the paradigm in medicine over the next decade. Neither genetic nor environmental explanations alone will be adequate any longer as mechanisms for human pathology. A remarkable variety of human diseases now appear to have a genetic susceptibility factor (see table 6.1). The interaction of the specific gene or its product with the environment is necessary to produce symptoms of the actual clinical disease. Although this level of understanding of pathology seemed impractical in the past, it has long been suggested as the ideal by workers in both biology and medicine.

CONSTITUTIONAL MEDICINE AND THE NEW GENETICS

In the field of human biology, attempts to integrate the genetic endowment with environmental influences have faltered over the years. It has always been easier to focus on one or the other issue. But the necessity to consider both aspects of this interaction was emphasized long ago in a neat summary by H. S. Jennings:

> All characteristics, then, are hereditary, and all are environmental. Does this deprive the study of the distinctive parts played by the two of all sense and value? It does not. It is of the greatest importance to know in what different ways diverse stocks respond effectively in the same environment; and how these diversities are perpetuated; what limitations the original constitution puts on what the environment can bring out; this is the study of heredity. It is equally important to know what differences appear among stock of the same original constitution under diverse environments; how great the possibilities of environmental actions are within a given stock. In man, where practically every individual represents a different stock and a different environment, the matter is not one of sweeping generalizations based on general biological principles. The concepts of the heredity and the environment cannot be employed in the absolute way now practiced; but they can be used with entire precision if

they are applied, not to characters-in-themselves, but to diversities between different particular cases. Though stature is always dependent on both heredity and environment, the difference in stature between Mr. Jones and Mr. Smith may be purely a matter of heredity; the difference between the same Mr. Jones and Mr. Brown may be purely a matter of environment. If there is clarity as to what comparison is made, there need be no ambiguity as to what is due to heredity, what to environment. (Jennings, 1924)

The small group of physicians who studied "constitutional medicine" around 1920 in the United States viewed human disease as a struggle between external factors from the environment and the inherent genetic endowment of each person. They believed that endogenous or "constitutional" factors were more important in causing disease than exogenous ones. George Draper, the leader of this school of thought in U.S. medicine, defined one's constitution as "that aggregate of hereditarial characters, influenced more or less by environment, which determines the individual's reaction, successful or unsuccessful, to the stresses of the environment." His work attempted to define physical types that he thought would reflect similar genetic predisposition. These types would then exhibit similar reactions to environmental triggers such as infections. In essence, Draper sought to integrate anatomy, physiology, immunology, and psychology to define the uniqueness of each person.

Draper believed that this detailed study would define the human types that were susceptible to insult from the environment. As early as 1916 he had observed that many victims of a polio epidemic shared several physical features: long eyelashes, dark hair, and widely spaced eyes. In that same year he established a Constitution Clinic at Presbyterian Hospital in New York, and for many years thereafter he investigated relationships between physical features ("types") and important human diseases, such as migraine headaches and peptic ulcers (Tracy, 1992).

A few other physicians became interested in constitutional medicine because it offered a pathological model distinct from the generally accepted exogenous theory, which claimed that virtually all human disease was the result of toxicants and germs. But as the success of the germ theory applied to public health problems such as tuberculosis and venereal disease was unmistakable, most U.S. physicians came to accept the exogenous model without much questioning as the most useful for daily clinical work.

By the end of his career in 1945, Draper was discouraged by the lack of interest in his interactional model for the cause of human disease. He noted that the key to better understanding of human constitution must "stem from the human gene, whose mysteries will form the next challenge" (Tracy, 1992).

CONCLUSION

The application of human genetic principles to specific cases has always been problematic because of the limited amount of precise knowledge of the human genome. The earliest workers in this field cautioned against the overzealous attempts of the eugenicists to simplify the complex human situation. Castle believed that only harm could result from attempts to magnify the available knowledge in an attempt to conceal its actual limitations (Castle, 1916).

In the near future, however, the information generated by the Human Genome Project will provide new opportunity for scientists and physicians to examine in great detail the genetic basis of many human diseases. Communication between the basic scientist and the medical practitioner will become ever more crucial if these new data are to be applied to actual human cases. "A basic understanding of molecular genetics is a necessity for physicians today . . . Many practicing physicians did not study genetics in their medical school curriculum. Continuing education will have to make up the clinical gaps in genetics and in DNA-based diagnostics" (Caskey, 1991).

It is important that this dialogue between the scientist and the physician enhance the goals that each is trying to achieve. The researcher seeks to better understand mechanisms of disease, while the practitioner attempts to apply this material to diagnose and treat patients afflicted by disease. As was the case one hundred years ago, this communication is once again fragmentary. J. D. Watson, one of the codiscoverers of the structure of DNA, has spoken rather disparagingly of the ability of the medical community to understand this new genetic revolution. He suggested that physicians were "probably not as appreciative as they should be, but that's asking an awful lot of them. Their lives are pretty full right now taking care of patients. They'll get excited when they have a reason to get excited—like the practical possibilities to make genetic diagnoses and to use genetic therapies" (Breo, 1989). In fact, physicians have already begun to prepare for the widespread application of this new genetic technology to clinical practice. Professional organizations such as the American Medical Association, the New York Academy of Medicine, and the Royal Society of Medicine have sponsored conferences to bring together research scientists and physicians to discuss new opportunities to understand the genetic basis of many important human diseases.

Major changes in medical paradigms have always been slow, as medicine by its nature is a conservative discipline. Physicians are practical people and continue to use ideas and technics that seem to work efficiently in the daily management of their sick patients. Austin Flint expressed this outlook nicely in 1876 when he noted that "the great event in the seventeenth century was the discovery of the circulation of the blood, in the eighteenth century the discovery of vaccination, and in the present century the discovery of anaesthesia. Events like this are not expected to recur at much shorter intervals" (Flint, 1876). Medical practitioners in the late nineteenth and early twentieth centuries also faced a radical change in scientific opinion on the role of genetic and environmental factors as causes for human disease. They concluded that most clinical disease was the result of environmental factors such as bacteria and toxicants, and not of hereditary factors.

Attempts by scientists and physicians to convince the medical community of the importance of genetic factors as causes for important human disease have proven largely unsuccessful, as this study has documented. In 1924 C. B. Davenport was appointed chair of a Committee on Heredity in Relation to Disease sponsored by the National Research Council. Several prominent physicians agreed to join this organization. L. Barker, C. Dana, G. Draper, L. Howe, and L. Hektoen met with Davenport periodically over five years and attempted to convince U.S. physicians that genetic factors should be studied as important causes of disease. Davenport believed that "we have reached the point both in the analysis of the factors of disease and defect, on the one hand, and the analysis of constitutional factors on the other to make it possible to undertake the cooperative work in the field of heredity by physicians and geneticists" (Tracy, 1992). But the time was not right, and the committee ended its efforts because few in medicine seemed to be listening to the message.

Modern physicians once again have the opportunity to use new genetic knowledge in their clinical work. The interaction of genetic and environmental factors will provide the model for understanding the mechanisms of disease in the twenty-first century. Conklin expressed the opinion many years ago that human genetic disorders were potential and not actual, that the outcome in each case always depended upon the interaction of heredity and the environment (Conklin, 1913). Physicians have the chance now more than ever to grasp the evolving information on the organization of the human genome and then to work with individual patients who may have altered genetic endowments. Conklin also believed that the bounds of human heredity were not narrow. He felt that within rather broad limits we humans had a considerable degree of both freedom and responsibility for our lives (Conklin, 1913). Physicians of the twenty-first century will be able to apply the new genetic paradigm as they strive with their patients to improve our capability to govern our own lives more effectively. This is *true* eugenics.

REFERENCES

Aaltonen, L. A., Peltomaki, P., and Leach, F. S. 1993. Clues to the pathogenesis of familial colorectal cancer. *Science* 260:812–815.

Abelmann, H. A. 1911. Heredity in cancer. *Illinois Medical Journal* 19:440–446.

Abt, I. A. 1911. Amaurotic family idiocy. *American Journal of Diseases of Children* 1:59–69.

Ackerknecht, E. H. 1982. Diathesis: The word and the concept in medical history. *Bulletin of the History of Medicine* 56:317–325.

Adami, J. G. 1901a. On theories of inheritance. *Montreal Medical Journal* 30:425–449.

———. 1901b. On theories of inheritance with special reference to the inheritance of acquired conditions in man. *New York Medical Journal* 73:925–936.

———. 1907. Inheritance and disease. In *Modern Medicine: Its Theory and Practice*, edited by W. Osler, 1:17–50. Philadelphia: Lea.

Adami, M. 1930. *J. George Adami: A Memoir.* London: Constable.

Adkinson, J. 1920. The behavior of bronchial asthma as an inherited character. *Genetics* 5:363–418.

Ager, L. C. 1919. The inheritance of acquired characters. *Journal of the American Medical Association* 73:1152.

Albert, D. M., and Scheie, H. G. 1965. *A History of Ophthalmology at the University of Pennsylvania.* Springfield, Ill.: Thomas.

Allen, F. M., and Mitchell, J. W. 1920. A case of hereditary diabetes. *Archives of Internal Medicine* 25:648–660.

Allen, G. E. 1968. T. H. Morgan and the problem of natural selection. *Journal of the History of Biology* 1:113–139.

———. 1983. The misuse of biological hierarchies: The American Eugenics Movement, 1900–1940. *History and Philosophy of the Life Sciences* 5:105–128.

———. 1986. The Eugenics Record Office at Cold Spring Harbor, 1910–1940: An essay in institutional history. *Osiris* 2:225–264.

Allen, N. 1869. The intermarriage of relations. *Quarterly Journal of Psychological Medicine* 3:244–297.

Aller, N. 1885–86. Laws of inheritance. *New England Medical Monthly* 5:197–201.

Alt, A. 1887. Some remarks on congenital cataracts. *American Journal of Pathology* 4:337–345.

Amick, W. R. 1884. Congenital absence of the pupil and malformation of the irides. *Cincinnati Lancet Clinic* 13:628–630.

Anderson, P. G. 1988. Personal communication, Medical Library, Washington University School of Medicine, St. Louis.

———. 1991. Personal communication, Medical Library, Washington University School of Medicine, St. Louis.

Andvist, J. W. 1913. Migraine. *St. Paul Medical Journal* 15:127–133.

Arps, G. F. 1920. Polydactylism. *Journal of the American Medical Association* 74:873–874.

Atkinson, J. E. 1875. Observation upon two cases of fibroma molluscum. *New York Medical Journal* 22:601–610.

Atwood, C. E. 1912. A brief report of three cases of family periodic paralysis. *New York State Journal of Medicine* 12:579–580.

Ayres, S. C. 1886. Retinitis pigmentosa. *American Journal of Ophthalmology* 3:81–90.

Badenhoop, K. 1990. Susceptibility to thyroid autoimmune disease. *Journal of Clinical Endocrinology and Metabolism* 71:1131–1137.

Baker, O. E. 1912. The eugenics club at the University of Wisconsin. *American Breeders Magazine* 3:69–71.

Ballantyne, J. W., and Elder, G. 1896. Tylosis palmae et plantae: With the description of two cases, mother and daughter. *Pediatrics* 1:337–349.

Ballentine, E. P. 1912. Review of six cases of hereditary chorea. *New York State Journal of Medicine* 12:644–647.

Baller, P. F. 1976. *American Thought in Transition: The Impact of Evolutionary Naturalism, 1865–1900*. Chicago: Rand McNally.

Bamberg, F. J. 1913. The prurigo of Hebra in five generations. *St. Paul Medical Journal* 15:514–516.

Barker, E. E. 1917. The present state of instruction in genetics. *Journal of Heredity* 8:69–71.

Barker, L. 1927. Heredity in the clinic. *American Journal of Medical Science* 173:597–605.

———. 1903. A description of the brains and spinal cords of two brothers dead of hereditary ataxia. *Transactions of the Association of American Physicians* 18:637–709.

———. 1942. *Time and the Physician: The Autobiography of Lewellys F. Barker*. New York: Putnam.

Barr, M. W. 1904. Heredity: Its influence for good or evil. *Alienist and Neurologist* 25:509–518.

Barss, H. de B. 1917. A report of two cases of von Recklinghausen's disease. *Journal of the Michigan State Medical Society* 16:113–116.

Barthalow, R. 1885. Diseases of the liver. In *System of Medicine*, edited by W. Pepper, 2:1033–1040. Philadelphia: Lea.

Bateson, W. 1900–1901. Problems of heredity as a subject for horticultural investigation. *Journal of the Royal Horticultural Society* 25:54–61.

———. 1902. Poultry. *Report to the Evolution Committee of the Royal Society* 1:87–124.

———. 1907. Facts limiting the theory of heredity. *Science* 26:649–660.

———. 1909. *Mendel's Principles of Heredity*. Cambridge: Cambridge University Press.

Bauer, H. H. 1992. *Scientific Literacy and the Myth of the Scientific Method*. Urbana: University of Illinois Press.

Bayley, W. D. 1897. Hereditary spastic paraplegia. *Journal of Nervous and Mental Diseases* 24:697–701.

Beard, G. M. 1880–81. Experiment with the "Jumpers of Maine." *Popular Science Monthly* 18:170–178.

Becker, O. 1880. Case of congenital unilateral colorblindness. *Boston Medical and Surgical Journal* 102:441–443.

Beecher, H. K. 1960. *Disease and the Advancement of Basic Science.* Cambridge, Mass.: Harvard University Press.

Bell, A. G. 1884. Memoir upon the formation of a deaf variety of the human race. *Memoirs of the National Academy of Science* 2:179–262.

———. 1907. A few thoughts concerning eugenics. *Proceedings of the American Breeders Association* 4:208–214.

———. 1909. Eugenics. *Proceedings of the American Breeders Association* 5:218–220.

Belt, E. O. 1896. Consanguinous marriage as a factor in the cause of disease. *Medical News* 68:61–64.

Bemiss, S. M. 1858. Report on influence of marriages of consanguinity upon offspring. *Transactions of the American Medical Association* 11:321–425.

Ben-David, J. 1960. Scientific productivity and academic organization in 19th century medicine. *American Sociological Review* 25:828–843.

Benedict, A. L. 1898. Heredity. *Medical Times* 26:193–196.

———. 1902. Heredity. *Medical Times* 30:289–291.

Berry, W. D. 1900. A contribution to the study of hereditary chorea. *American Journal of Insanity* 57:331–339.

Bidwell, R. I. 1911. Theories of etiology of cancer. *Ohio Medical Journal* 7:187–189.

Bigelow, J. 1854. *Nature in Disease.* Boston: Ticknor and Fields.

Billings, J. S. 1885. *Index Catalogue of the Library of the Surgeon General's Office.* Vol. 6. Washington, D.C.: Government Printing Office.

Bishop, J. E., and Waldholz, M. 1990. *Genome: The Story of the Most Astonishing Adventure of Our Time—The Attempt To Map All the Genes in the Human Body.* New York: Simon and Schuster.

Blacher, L. I. 1982. *The Problem of the Inheritance of Acquired Characters.* New Delhi: American.

Blackford, C. M. 1915–16. Some facts and problems of heredity. *Virginia Medical Semi-Monthly* 20:183–187.

Blackwell, L. S. 1903. Later impressions of the heredity of acquired characteristics. *Medical News* 83:500.

Blades, W. F. 1914. Heredity as a factor in congenital hare-lip and cleft palate. *Dental Cosmos* 56:1241–1245.

Bluhm, A. 1912. *Problems in Eugenics.* London: Eugenics Education Society.

Blum, K., Noble, E. P., and Sheridan, P. J. 1990. Allelic association of human dopamine D2 receptor gene in alcoholism. *Journal of the American Medical Association* 263:2055–2060.

Bodmer, W. F. 1991a. Cancer genetics and the human genome. *Hospital Practice* 26:101–117.

———. 1991b. The Human Genome. *American Journal of Human Genetics* 49:1.

Bonner, T. N. 1963. *American Doctors and German Universities.* Lincoln: University of Nebraska Press.

Bordley, J. 1908. A family of hemerolopes. *Johns Hopkins Hospital Bulletin* 19:278–280.

Bovee, J. W. 1909. Large polycystic kidney. *American Journal of Obstetrics* 60:47–60.

Bowen, Professor. 1860–62. Heredity. *Proceedings of the American Academy of Arts and Sciences* 5:102–111.

Bowers, J. Z. 1976. Influences on the development of American medicine. In *Advances in American Medicine: Essays at the Bicentennial*, edited by J. Z. Bowers and E. F. Purcell, 1–38. New York: Macy Foundation.

Bowler, P. J. 1989. *The Mendelian Revolution.* Baltimore: Johns Hopkins University Press.

Boyd, W. A. 1913. Hereditary chorea. *Boston Medical and Surgical Journal* 169:680–683.

Braishin, W. C. 1904. Microtia with an account of two cases occurring in members of the same family. *Brooklyn Medical Journal* 18:136–137.

Brandeis, J. W. 1915. Polydactylism as a hereditary character. *Journal of the American Medical Association* 64:1640–1642.

———. 1918. A note on amaurotic family idiocy. *New York Medical Journal* 107:121.

Brandy, J. L. 1960. Neurology and Psychiatry. In *The Education of American Physicians*, edited by R. L. Numbers, 226–245. Berkeley: University of California Press.

Breo, D. L. 1989. DNA discoverer J. D. Watson now dreams of curing genetic diseases. *Journal of the American Medical Association* 262:3340–3344.

Bridge, N. 1885. Pseuodohypertrophic muscular paralysis. *Medical News* 46:29–30.

Brieger, G. H. 1983. "Fit to study medicine": Notes for a history of premedical education in America. *Bulletin of the History of Medicine* 57:1–21.

Briggs, H. H. 1918. Hereditary congenital ptosis: With report of 64 cases conforming to the Mendelian rule of dominance. *Transactions of the American Ophthalmological Society* 16:255–276.

Briggs, J. R. 1885–86. Congenital absence of external auditory meatus, the result of maternal impressions. *Texas Courier-Record of Medicine* 3:95–96.

Bristol, L. D. 1916. An enzyme theory of cancer etiology. *Medical Record* 89:180–191.

Bromwell, B. 1902–3a. Case of hereditary ichthyosis of the palms and soles. *Clinical Studies* 1:77–80.

———. 1902–3b. Ichthyosis. *Clinical Studies* 1:74–77.

Brooks, H. 1909. Three cases of amaurotic family idiocy. *Transactions of the Association of American Physicians* 24:12–19.

Brophy, T. W. 1901. Surgical treatment of palatal defects. *Dental Cosmos* 23:317–340.

Brower, D. R. 1897. Hereditary ataxia: Friedreich's disease. *Journal of the American Medical Association* 28:871–872.

Brown, C. W. 1873–74. Supernumary fingers and toes. *Medical Times* 4:278.

Brown, J. 1845. Hereditary transmission of disease. *Cyclopedia of Practical Medicine* 2:443–445.

Brown, S. 1891–92. A report of a series of twenty-one cases of hereditary ataxy. *Chicago Medical Recorder* 2:500–516.

Brown-Sequard, C. E. 1875. On the hereditary transmission of effects of certain injuries to the nervous system. *Lancet* 1:7–8.

Broyton, A. W. 1892–93. Xeroderma pigmentosum: Three cases in one family. *Indiana Medical Journal* 11:65–68.

Bruce, R. V. 1987. *The Launching of Modern American Science.* New York: Knopf.

Bruner, W. E. 1912. Hereditary optic atrophy. *Transactions of the American Ophthalmological Society* 13:162–174.

Brush, A. C. 1895. Huntington's chorea. *New York Medical Journal* 61:305–306.

Bryant, D. C. 1892. Lens dislocation. *Journal of the American Medical Association* 19:278.

Buchanan, J. A. 1920. The Mendelism of migraine. *Medical Record* 98:807–808.

———. 1921. The familial distribution of the migraine-epilepsy syndrome. *New York Medical Journal* 113:45–47.

Buchanan, M. 1907. A case of amaurotic family idiocy. *Annals of Ophthalmology* 16:249–255.

Buel, S. 1817. An account of a family predisposition to hemorrhage. *Transactions of the Physico-Medical Society of New York* 1:305–310.

Bullard, F. D. 1890. Paternal impressions. *Southern California Practitioner* 5:324–327.

Bunting, C. H. 1906. Congenital cystic kidney and liver with family tendency. *Journal of Experimental Medicine* 8:271–288.

Burch, M. G. 1912. A case of hemophilia. *Medical Times* 40:324–325.

Burnett, S. G. 1912. A clinical study of the family tree transcendency of migraine. *Medical Herald* 39:342–350.

Burnett, S. M. 1882. Color blindness and color perception. *Popular Science Monthly* 21:86–96.

Burnett, S. W. 1879. Results of an examination of the color sense of 3,040 children. *Archives of Ophthalmology* 8:191–199.

Burr, C. W. 1893–94. A contribution to the pathology of Friedreich's ataxia. *University Medical Magazine* 6:598–604.

———. 1917. Heredity in St. Vitus dance. *Journal of Nervous and Mental Diseases* 45:237–240.

———. 1922. Heredity in epilepsy. *Archives of Neurology* 7:721–728.

Butler, W. M. 1894. Huntington's chorea with report of two additional cases. *North American Journal of Homeopathy* 9:139–143.

Buxton, L. H. 1916. A dominant Mendelian inheritance, principal factor in occurrence of cataract and ectopia lentis. *Southern Medical Journal* 9:933–938.

Byford, H. T. 1915. Notes on the etiology and prophylaxis of cancer. *Transactions of the Western Surgical Association*, 233–246.

Byford, W. H. 1886. Carcinoma or cancer of the uterus. In *A System of Practical Medicine*, edited by W. Pepper, 4:274–275. Philadelphia: Lea.

Calhoun, F. P. 1914. Hereditary glaucoma. *Journal of the American Medical Association* 63:209–215.

Calvin, W. D. 1908. Etiology of hare-lip and cleft palate. *Dental Digest* 14:1508–1551.

Camac, C. N. B. 1915. Hemophilia. *A Reference Handbook in the Medical Sciences*, edited by T. L. Stedman, 5:176–178. New York: Wood.

Camp, C. D. 1916. A case of pseudohypertrophic muscular dystrophy. *Journal of the Michigan Medical Society* 15:317.

Campbell, J. A. 1913. Hereditary cataract. *Journal of Ophthalmology, Otology, and Laryngology* 19:144–147.

Cannon, W. A. 1902. A cytological basis for the Mendelian laws. *Bulletin of the Torrey Botanical Club* 29:657–661.

Carlow, C. M. 1901. Heredity from a medical standpoint. *Medical Dial* 3:49–56.

Carlson, E. A. 1966. *The Gene: A Critical History.* Philadelphia: Saunders.

Carrington, H. A. 1868–69. Heredity. *Proceedings of the Connecticut Medical Society* 3:365–376.

Carter, C. B. 1894. A case of rare and fatal disease of infancy with symmetrical changes in the yellow spot. *Archives of Ophthalmology* 23:126–130.

Caskey, C. T. 1991. Physician-laboratory interface in X chromosome mapping. *Hospital Practice* 26:131–144.

Cassell, E. J. 1979. Changing ideas of causality in medicine. *Social Research* 46:728–743.

Castle, W. E. 1902–3. Mendel's law of heredity. *Proceedings of the American Academy of Arts and Sciences* 38:533–548.

———. 1903a. Heredity of sex. *Bulletin of the Museum of Comparative Zoology* 40:189–219.

———. 1903b. The heredity of agora coat in mammals. *Science* 18:760.

———. 1903c. The laws of heredity of Galton and Mendel, and some laws governing race improvement. *Proceedings of the American Academy of Arts and Sciences* 39:223–242.

———. 1903d. Mendel's law of heredity. *Science* 18:396–406.

———. 1903e. Notes on Mr. Farabee's observations. *Science* 17:75–76.

———. 1909. On some laws of inheritance. *Illinois Medical Journal* 15:406–412.

———. 1916. *Genetics and Eugenics.* Cambridge, Mass.: Harvard University Press.

———. 1951. The beginnings of Mendelism in America. In *Genetics in the 20th Century*, edited by L. C. Dunn, 59–76. New York: Macmillan.

Castle, W. E., and Allen, G. M. 1903. The heredity of albinism. *Proceedings of the American Academy of Arts and Sciences* 38:603–622.

Castle, W. E., and Phillips, J. C. 1909. A successful ovarian transplantation in the guinea pig and its bearing on problems of heredity. *Science* 30:312–313.

Cathell, D. W. 1882. *The Physician Himself, and What He Should Add to His Scientific Acquirements.* Baltimore: Cushings and Bailey.

Chancellor, C. W. 1887. Heredity and other peculiarities. *Annual Report of the Pennsylvania State Board of Health* 2:234–242.

Chase, R. H. 1907. Morbid inheritance. *Medical Times* 35:161–162.

Cheatham, W. 1903. Cataract in the young. *Medical Age* 21:681–683.

Child, W. 1890. Inheritance of disease. *Transactions of the New Hampshire Medical Society*, 41–48.

Christian, E. P. 1889. Philosophy of causation of some congenital abnormalities of structure. *American Lancet* 13:41–45.

Christison, J. S. 1895. Inherited peculiarities. *Journal of the American Medical Association* 25:1040–1042.

Church, A. 1906. The neuritic type of progressive muscular atrophy: A case with marked heredity. *Journal of Nervous and Mental Diseases* 33:447–453.

Church, B. F. 1914. The Mendelian law and its relation to inherited conditions of the eye. *California State Journal of Medicine* 12:506–507.

Clark, C. P. 1912. Remarks upon some recent studies in the pathogenesis of epilepsy. *Boston Medical and Surgical Journal* 167:78–81.

Clark, J. H. 1880–81. Heredity. *Ohio Medical Recorder* 5:407–416.

Clark, L. P. 1915. Epilepsy. In *Modern Medicine*, edited by W. Osler and T. McCrae, 5:592–623. Philadelphia: Lea and Febiger.

Claus, E. B., Risch, N., and Thompson, W. D. 1991. Genetic analysis of breast cancer in the cancer and steroid hormone study. *American Journal of Human Genetics* 48:232–242.

Clemesha, J. C. 1897–98. Thomsen's disease: A family history. *Buffalo Medical Journal* 37:16–18.

Cleveland, J. L. 1877. Consumption and heredity. *Cincinnati Clinic* 13:1–3.

Clowe, C. F. 1909. Mind and its relation to heredity. *New York Medical Journal* 90:391–395.

Coe, A. H. 1888. Case of supernumary nipple associated with maternal impressions. *Medical Record* 34:479.

Coe, H. W. 1896. Thomsen's disease—Friedreich's ataxia—Acromegaly. *Medical Sentinel* 4:415–423.

Cohen, M., and Dixon, G. S. 1907. Report of a case of amaurotic family idiocy. *Journal of the American Medical Association* 48:1751–1753.

Coleman, F. W. 1889. Retinitis pigmentosa with special reference to the question whether it affords evidence of degeneration from the marriage of near kin. *North American Practitioner* 1:49–58.

Collins, J. 1898a. The pathology and morbid anatomy of Huntington's chorea. *American Journal of Medical Science* 116:275–291.

———. 1898b. The pathology and morbid anatomy of Huntington's chorea. *Journal of Nervous and Mental Diseases* 25:57–61.

———. 1903. A clinical report of nine cases of Friedreich's disease: Hereditary or family ataxia. *American Medicine* 5:865–870.

———. 1905. Hereditary progressive muscular atrophy. *The Post-Graduate* 20:510–516.

Conklin, E. G. 1908. The mechanism of heredity. *Science* 27:89–99.

———. 1913. Heredity and responsibility. *Science* 37:46–54.

Connolly, J. M. 1912. Heredity: With special reference to the law of Gregor Mendel. *Boston Medical and Surgical Journal* 167:791–798, 836–844.

Cook, O. F. 1903. Evolution, cytology, and Mendel's laws. *Popular Science Monthly* 63:219–228.

Coriat, I. H. 1913. Amaurotic family idiocy. *Archives of Pediatrics* 30:404–415.

———. 1916. Dystonia musculorum deformans. *Boston Medical and Surgical Journal* 175:383–386.

Correns, C. 1900. G. Mendel's Regel uber das Verhalten der Nachkommenschaft der Rassenbastarde. *Berichte der Deutscher botanischer Gesellschaft* 18:158–168.

Cottral, G. H. 1905. Huntington's chorea. *St. Louis Courier of Medicine and Collateral Sciences* 33:17–23.

Couch, A. S. 1880–81. Heredity and the higher duties of the profession. *Transactions of the Homeopathic Medical Society of New York* 16:122–134.

Coulter, J. W. 1896. Heredity. *Medical Examiner* 6:226.

Councilman, W. T. 1913. The nature of disease. *California State Journal of Medicine* 11:260–266.

Cowie, D. M. 1913. Hypodactylism. *Physician and Surgeon* 25:131.

Craig, C. B. 1916. A case of hereditary tremor. *Boston Medical and Surgical Journal* 174:507–508.

Crandall, F. M. 1897. Heredity and degeneration. *Archives of Pediatrics* 14:894–900.

Cravens, H. 1978. *The Triumph of Evolution: American Scientists and the Heredity-Environment Controversy.* Philadelphia: University of Pennsylvania Press.

Crowder, J. R., and Crowder, T. R. 1917. Angioneurotic edema in 5 generations. *Archives of Internal Medicine* 20:840–852.

Crowell, H. C. 1900. Improve the species. *Journal of the American Medical Association* 34:329–331.

A Curious case of albinism. 1884. *Popular Science Monthly* 25:122.

Cushing, H. 1916a. Hereditary ankylosis of the proximal phalangeal joints. *Journal of Nervous and Mental Diseases* 43:445–446.

———. 1916b. Hereditary ankylosis of the proximal phalangeal joints (symphalangism). *Genetics* 1:90–106.

Dana, C. L. 1887. Hereditary tremor. *American Journal of Medical Science* 94:386–393.

———. 1895. The pathology of hereditary chorea. *Journal of Nervous and Mental Diseases* 22:565–583.

———. 1903. A case of choreic tic. *Boston Medical and Surgical Journal* 148:449–451.

———. 1907. A family type of combined sclerosis associated with grave anemia. *Medical Record* 72:1081.

———. 1910. The modern views of heredity. *Medical Record* 77:345–350.

Danforth, C. H. 1914. Some notes on a family with hereditary congenital cataract. *American Journal of Ophthalmology* 31:161–172.

———. 1916. Some aspects of the study of hereditary eye defects. *American Journal of Ophthalmology* 33:65–70.

Danforth, G. 1888–89. A law of heredity or possibly maternal impressions. *Texas Courier-Record of Medicine* 6:79–81.

Darden, L. 1991. *Theory Change in Science: Strategies from Mendelian Genetics.* New York: Oxford University Press.

Darwin, C. 1868. *The Variation of Animals and Plants under Domestication.* Vol. 2. London: John Murray.

———. 1898. *The Variation of Plants and Animals under Domestication.* 2d ed. New York: Appleton.

Davenport, C. B. 1901. Mendel's law of dichotomy in hybrids. *Biological Bulletin* 2:307–310.

———. 1904. Wonder horses and Mendelism. *Science* 19:151–153.

———. 1907. Recent advances in the theory of breeding. *Proceedings of the American Breeders Association* 3:131–135.

———. 1907–8. Heredity of some human physical characteristics. *Proceedings of the Society of Experimental Biology and Medicine* 5:101–102.

———. 1909. Influence of heredity in human society. *Annals of the American Academy of Political and Social Sciences* 34:16–21.

———. 1910a. Eugenics, a subject for investigation rather than instruction. *American Breeders Magazine* 1:68–69.

———. 1910b. Fit and unfit matings. *Bulletin of the American Academy of Medicine* 11:657–670.

———. 1910c. Heredity in man. In *The Harvey Lectures, 1908–1909*, 280–290.

———. 1910d. Report of Committee on Eugenics. *American Breeders Magazine* 1:126–129.

———. 1910–11. Application of Mendel's law to human heredity. *Journal of Asthenics* 15:93–95.

———. 1911. *Heredity in Relation to Eugenics*, New York: Holt.

———. 1912a. Heredity in nervous disease and its social bearing. *Journal of the American Medical Association* 59:2141–2142.

———. 1912b. Some practical lessons for neurologists drawn from recent eugenic studies. *Journal of Nervous and Mental Diseases* 39:402–405.

———. 1912c. Some social applications of modern principles of heredity. *Transactions of the 15th International Congress of Hygiene and Demography* 4:658–662.

———. 1913. Inheritance of some of the elements of hysteria. *Illinois Medical Journal* 24:289–290.

———. 1914. Heredity of some emotional traits. *Science* 39:567–568.

———. 1915a. The feebly inherited: I. Violent temper and its inheritance. *Journal of Nervous and Mental Diseases* 42:593–628.

———. 1915b. Huntington's chorea in relation to heredity and eugenics. *Proceedings of the National Academy of Sciences* 1:283–285.

———. 1915c. Inheritance of Huntington's chorea. *Science* 41:570–571.

———. 1918. Hereditary tendency to form nerve tumors. *Proceedings of the National Academy of Sciences* 4:213–214.

———. 1921. Research in eugenics. *Science* 54:391–397.

———. 1930. Sex linkage in man. *Genetics* 15:401–444.

———. 1936. Letter to Simon Flexner regarding biography of William Welch, May 12, 1936. A. M. Chesney Medical Archive, Johns Hopkins Medical Institutions, Baltimore.

Davenport, C. B., and Davenport, G. C. 1908. Heredity in hair form in man. *American Naturalist* 42:341–349.

Davenport, C. B., and Muncey, E. B. 1916. Huntington's chorea in relation to heredity and eugenics. *American Journal of Insanity* 73:195–222.

Davenport, C. B., and Weeks, D. F. 1911. A first study of inheritance of epilepsy. *Journal of Nervous and Mental Diseases* 38:641–670.

Davenport, G. C., and Davenport, C. B. 1907. Heredity of eye color in man. *Science* 26:589–592.

———. 1909. Heredity of hair color in man. *American Naturalist* 43:193–211.

———. 1910. Heredity of skin pigment in man. *American Naturalist* 44:642–672, 705–731.

Davidson, R. L., and Childs, B. 1987. Perspectives on the teaching of human genetics. *Advances in Human Genetics* 16:79–119.

Davis, A. E. 1909. Amaurotic family idiocy. *Post-Graduate* 24:251–263.

Dawkins, R. L., Martin, E., and Saueracker, G. 1990. Supratypes and ancestral haplotypes in IDDM. *Journal of Autoimmunity* 3:63–68.

Dean, L. W. 1903. Degenerate ocular changes resulting from consanguinity. *American Journal of Ophthalmology* 20:337–343.

DeBeck, D. 1886. A rare family history of congenital coloboma of the iris. *Archives of Ophthalmology* 15:8–23.

———. 1894. A family history of iridemia and coloboma irides. *Transactions of the American Ophthalmological Society*, 117–128.

———. 1897. Retinitis pigmentosa. *Ohio Medical Journal* 8:235–236.

———. 1900. Family history of micro-cornea. *Cincinnati Lancet Clinic* 44:175–176.

DeCosse, J., Miller, H., and Lesser, M. 1989. Effect of wheat fiber and vitamin C and E on rectal polyps in patients with familial adenomatous polyposis. *Journal of the National Cancer Institute* 81:1290–1297.

DeFontenay, O. E. 1881. Results of examination for color blindness in Denmark. *Archives of Ophthalmology* 10:8–19.

DeJong, R. N. 1965. The founding of the American Neurological Association. *Transactions of the American Neurological Association* 90:3–11.

———. 1982. *A History of American Neurology*. New York: Raven.

Dercum, F. X. 1915. Nervous and mental diseases and nerve pathology. *Journal of Nervous and Mental Diseases* 42:358–369.

DeVries, H. 1889. *Intracellulare Pangenesis*. Jena: Fischer.

———. 1900. Das Spaltungsgesetz der Bastarde. *Berichte der Deutscher botanischer Gesellschaft* 18:83–90.

Dickey, J. L. 1898. Another cataractous family. *Journal of the American Medical Association* 31:997.

Diller, T. 1889. Some observations on the hereditary form of chorea. *American Journal of Medical Science* 98:585–593.

Dixon, J. T. 1912–13. Heredity and its influences on progeny. *Kentucky Medical Journal* 11:635–638.

Dodge, A. H. 1906. An isolated case of Friedreich's ataxia. *Journal of the American Medical Association* 46:802–803.

Dolan, T. M. 1888–89. Heredity. *New England Medical Monthly* 8:16–21, 54–60, 97–103, 149–157.

Doran, R. E. 1903–4. A consideration of the hereditary factors in epilepsy. *American Journal of Insanity* 60:61–73.

Dronamraju, K. R. 1989. *The Foundations of Human Genetics*. Springfield, Ill.: Thomas.

Dryja, T. P., Hahn, L. B., and McGee, T. L. 1991. Mutation spectrum of the rhodopsin gene among patients with autosomal dominant retinitis pigmentosa. *American Journal of Human Genetics* 49:2.

Duhring, L. A., and Stelwagon, H. W. 1886. Diseases of the skin. In *A System of Practical Medicine*, edited by W. Pepper, 4:583–733. Philadelphia: Lea.

Duke-Elder, S. 1958. *A Century of International Ophthalmology, 1857–1957*. London: Kimpton.

Dupuy, E. 1877. On heredity in nervous disorders. *Popular Science Monthly* 11:332–339.

Earle, P. 1845. On the inability to distinguish colors. *American Journal of Medical Science* 9:346–354.

East, E. M. 1922. As genetics comes of age. *Journal of Heredity* 13:207–214.

Ehrenfield, A. 1917. Hereditary deforming chondrodysplasia: More cases. *American Journal of Orthopedic Surgery* 15:463–478.

Eisenstaedt, J. S. 1913. Three cases of family dystrophy of the hair and nails. *Journal of the American Medical Association* 60:27–28.

El-Deiry, W. S., Nelkin, B. D., and Celano, P. 1991. High expression of the DNA methyl-transferase gene characterizes human neoplastic cells and progression stages of colon cancer. *Proceedings of the National Academy of Sciences* 88:3470–3474.

Elliot, G. T. 1900. A contribution to the histopathology of epidermolysis bullosa. *New York Medical Journal* 71:585–588, 625–629.

Elsberg, L. 1879. On the explanation of hereditary transmission. *Proceedings of the American Association for the Advancement of Science* 28:520–523.

———. 1882. Changes in biological doctrines during the past 25 years. *College and Clinical Record* 3:71–77.

Engman, M. F. 1913. A psoriasis family tree. *Journal of Cutaneous Diseases* 31:559–560.

Engman, M. F., and Mook, W. H. 1906. A study of some cases of epidermolysis bullosa. *Journal of Cutaneous Diseases* 24:55–67.

———. 1910. The etiology of epidermolysis bullosa. *Interstate Medical Journal* 17:499–501.

Epstein, J. 1917a. Amaurotic family idiocy. *New York Medical Journal* 106:887–889.

———. 1917b. The diatheses in childhood. *Medical Record* 91:1132–1133.

———. 1920. Amaurotic family idiocy. *Medical Record* 97:224–227.

Erlich, H. A., Bugawan, T. L., and Scharf, S. 1990. HLA-DQ beta sequence polymorphism and genetic susceptibility to IDDM. *Diabetes* 39:96–103.

Eshner, A. A. 1896–97. Hereditary lateral sclerosis. *University Medical Magazine* 9:809–818.

Eugenics in the colleges. 1914. *Journal of Heredity* 5:186.

Fabio, G., Smeraldi, R. S., and Girgeri, A. 1990. Susceptibility to HIV infection and AIDS in Italian hemophiliacs is HLA associated. *British Journal of Hematology* 75:531–536.

Fairbanks, A. W. 1904. Hereditary oedema. *American Journal of Medical Science* 127:877–891.

———. 1914. A study of the etiology of one hundred and seventy five epileptic children. *Boston Medical and Surgical Journal* 170:521–523.

Fairchild, D. 1920. Physicians and genetics. *Journal of the American Medical Association* 74:48.

Farabee, W. C. 1903. Notes on Negro albinism. *Science* 17:75.

———. 1905. Inheritance of digital malformations in man. *Peabody Museum of American Archaeology and Ethnology Papers* 3:67–80.

———. 1912. Cephalic type contours. *Science* 35:673.

———. 1918. The Arawaks of northern Brazil. *American Journal of Physical Anthropology* 1:427–442.

Farrall, L. A. 1985. *The Origins and Growth of the English Eugenics Movement, 1865–1925.* New York: Garland.

Fay, E. A. 1898. *Marriages of the Deaf in America.* Washington, D.C.: Volta Bureau.

Fearon, E. R., Cho, K. R., and Nigro, J. M. 1990. Identification of a chromo-some 18q gene that is altered in colorectal cancers. *Science* 247:49–56.

Feeblemindedness. 1915. *Journal of Heredity* 6:31–36.

The Feebly Inhibited: Nomadism or the wandering impulse. 1915. New York: Carnegie Institution.

Feingold, M. 1916. Progressive macular degeneration in three members of one family. *Archives of Ophthalmology* 45:533–543.

Fellows, H. B. 1886. Friedreich's disease, or hereditary ataxy. *Clinique* 7:1–8.

Festinger, L. 1957. *A Theory of Cognitive Dissonance.* Stanford, Calif.: Stanford University Press.

Finlayson, H. W. 1916. *The Dack Family: A Study in Hereditary Lack of Emotional Control.* Cold Spring Harbor, N.Y.: Eugenics Record Office.

Fischer, J. C. 1914. Inheritance with reference to the eye and ear. *Illinois Medical Journal* 26:499–501.

Fitz, R. H. 1885. General morbid processes. In *A System of Practical Medicine,* edited by W. Pepper, 1:35–124. Philadelphia: Lea.

Fleming, D. 1954. *William H. Welch and the Rise of Modern Medicine.* Boston: Little, Brown.

Flemming, W. 1882. *Zellsubstanz, Kern, und Zelltheilung.* Leipzig: Vogel.

Flexner, A. 1910. *Medical Education in the United States and Canada.* New York: Carnegie Foundation.

Flexner, S., and Flexner, J. T. 1941. *William Henry Welch and the Heroic Age of American Medicine.* New York: Viking.

Flint, A. 1876. The first century of the republic: Medical and sanitary progress. *Harper's New Monthly Magazine* 53:83.

Fol, H. 1877. On the beginning of ontogeny in various animals. *Archives de Zoologie experimentale et generale* 6:145–169.

Fox, G. H. 1884. The "Alligator Boy": A case of ichthyosis. *Journal of Cutaneous and Venereal Disease* 2:97–99.

Fox, L. W. 1882. Examination of Indians at the government school in Carlisle, PA for acuteness of vision and colorblindness. *Philadelphia Medical Times* 12:346–347.

Franek, Z., Timmerman, L. A., and Alper, C. A. 1990. Major histocompatability complex genes and susceptibility to SLE. *Arthritis and Rheumatology* 33:1542–1553.

Frank, I. 1903. The hereditary influence in nystagmus is remarkably illustrated. *Medical Record* 63:175.

Frank, M. 1903–4. Hereditary tendency to refractive errors in a family. *Illinois Medical Journal* 5:329–330.

———. 1906. Amaurotic family idiocy. *Pediatrics* 18:148–151.

Franklin, C. P. 1913. Eugenics from the point of view of the ophthalmologist. *Journal of the Medical Society of New Jersey* 7:437–439.

Frazier, B. C. 1903–4. Hereditary Friedreich's ataxia. *Louisville Monthly Journal of Medicine and Surgery* 10:191–192.

French, M. A., and Dawkins, R. L. 1990. Central MHC genes, IgA deficiency and autoimmune disease. *Immunology Today* 11:271–274.

Fry, F. E. 1896. Additional cases of Friedreich's ataxia. *Medical Review* 34:19–22.

Fry, F. R. 1893. Friedreich's ataxia: Two cases. *Medical Fortnightly* 4:638–649.

Fuller, E. M. 1887. Pedigree in health and disease. *Transactions of the Maine Medical Association*, 189–210.

Furnas, J. C. 1969. *The Americans: A Social History*. New York: Putnam.

Gaddy, N. D. 1883. Heredity. *Medical Herald* 5:241–249.

Galton, F. 1871. Experiments on pangenesis. *Proceedings of the Royal Society, London* 19:393–410.

———. 1876. A theory of heredity. *Journal of the Anthropological Institute of Great Britain and Ireland* 5:329–348.

———. 1883. *Inquiries into Human Faculty and Its Development*. London: Macmillan.

Gardner, W. A. 1913–14. Heredity in nervous disease and its social bearing. *Atlanta Journal and Record of Medicine* 60:456–461.

Garrod, A. E. 1902. The incidence of alkaptonuria. *Lancet* 2:1616–1620.

Garver, J. K. 1895. Heredity. *Transactions of the Medical Society of Pennsylvania* 26:276–280.

Gasser, H. 1895. The dynamics of heredity. *Medical Record* 47:673–677.

Geneology and eugenics. 1915. *Journal of Heredity* 6:379.

Gibney, V. P. 1876. Hereditary multiple exostoses. *American Journal of Medical Science* 72:73–80.

Gilbert, W. 1991. Towards a paradigm shift in biology. *Nature* 349:99.

Gilchrist, T. C. 1897. Eleven cases of parakeratosis in one family. *Journal of Cutaneous and Genito-Urinary Diseases* 17:149–172.

Gillispie, C. C. 1976. *Dictionary of Scientific Biography*. New York: Scribners.

Glass, B. 1968. Maupertuis, pioneer of genetics and evolution. In *Forerunners of Darwin, 1745–1859*, edited by B. Glass, O. Temkin, and W. L. Straus, 51–83. Baltimore: Johns Hopkins University Press.

———. 1986. Geneticists Embattled: Their stand against rampant eugenics and racism in America during the 1920s and 1930s. *Proceedings of the American Philosophical Society* 130:130–154.

Goddard, H. H. 1912. Heredity of feeble-mindedness. *Proceedings of the American Philosophical Society* 51:173–177.

Gordan, W. S. 1897. Heredity as a causative factor in disease. *Practice* 11:111–118.

Gordon, A. 1915. Mendelian laws of heredity and their relation to eugenics. *Virginia Medical Semi-Monthly* 20:53–60.

Gould, G. M. 1893a. Homeochronous hereditary optic atrophy. *American Journal of Ophthalmology* 10:278–279.

———. 1893b. Homeochronous hereditary optic atrophy, extending through six generations. *Annals of Ophthalmology and Otology* 2:303–307.

———. 1893c. Homeochronous hereditary optic nerve atrophy, extending through six generations. *Transactions of the Pan-American Medical Congress*, 1387.

Graham, J. E. 1885. A case of tuberculo-ulcerative syphilide of hereditary origin. *Medical News* 47:273.

Grannan, St. G. J. 1910. Hereditary ataxia. *Pediatrics* 22:231–234.

Graves, W. P. 1910. Present knowledge of the laws of heredity. *Boston Medical and Surgical Journal* 163:829–838.

Gray, L. C. 1879. A case of extraordinary heredity in epilepsy. *Archives of Medicine* 1:215.

Green, E. D., and Waterston, R. H. 1991. The Human Genome Project: Prospects and implications for clinical practice. *Journal of the American Medical Association* 266:1966–1975.

Griffin, H. Z., and Sanford, A. H. 1919. Clinical observations regarding the fragility of erythrocytes. *Journal of Laboratory and Clinical Medicine* 4:465–478.

Griffith, F. 1910. Case of congenital fusion of the toes with note of previous generations. *Medical Record* 78:67.

Griffith, J. P. C. 1888. A contribution to the study of Friedreich's ataxia. *American Journal of Medical Science* 96:377–388.

Griswold, R. M. 1881. Some observations upon heredity: Tendencies and transmissions. *Independent Practitioner* 2:515–522, 587–591.

Guilford, S. H. 1883. A dental anomaly. *Dental Cosmos* 25:113–118.

Guyer, M. F. 1900. Spermatogenesis in hybrid pigeons. *Science* 11:248–249.

———. 1902a. Hybridism and the germ cell. *Bulletin of the University of Cincinnati* 2:1.

———. 1902b. Spermatogenesis of normal and hybrid pigeons. *Bulletin of the University of Cincinnati* 3:1.

———. 1903. The germ cell and the results of Mendel. *Cincinnati Lancet Clinic* 1:490–491.

———. 1907. Do offspring inherit equally from each parent? *Science* 25:1006–1010.

———. 1909. Deficiencies of the chromosome theory of heredity. *University Studies* 5:3–17.

———. 1910. Accessory chromosomes in man. *Biological Bulletin* 19:219–234.

Haeckel, E. 1866. *Generalle Morphologie der Organismen.* Vol. 2. Berlin: Reinner.

Hagedoorn, L., and Hagedoorn, A. L. 1920. Inherited predisposition for a bacterial disease. *American Naturalist* 54:368–375.

Halbert, H. V. 1895. Two cases of Friedreich's ataxia. *Clinique* 16:636–640.

Hall, J. M., Friedman, C. and Guenther, C. 1992. Closing in on a breast cancer gene on chromosome 17q. *American Journal of Human Genetics* 50:1235–1242.

Hall, J. M., Lee, M. K., and Newman, B. 1990. Linkage of early-onset familial breast cancer to chromosome 17q21. *Science* 250:1684–1689.

Haller, J. S. 1981. *American Medicine in Transition, 1840–1910.* Urbana: University of Illinois Press.

Haller, M. H. 1984. *Eugenics: Hereditarian Attitudes in American Thought.* New Brunswick, N.J.: Rutgers University Press.

Hallock, F. K. 1898. A case of Huntington's chorea. *Journal of Nervous and Mental Diseases* 25:851–864.

Hamilton, A. J. 1908. A report of 27 cases of chronic progressive chorea. *American Journal of Insanity* 64:403–475.

Hamilton, A. M. 1886. Epilepsy. In *A System of Practical Medicine*, edited by W. Pepper, 5:467–503, Philadelphia: Lea.

Hamilton, A. S. 1916. A report of two cases of progressive lenticular degeneration. *Journal of Nervous and Mental Diseases* 43:297–323.

Hanes, F. M. 1909. Multiple hereditary telangiectasiases causing hemorrhage. *Johns Hopkins Hospital Bulletin* 20:63–73.

Hansell, H. F. 1895–96. A congenital cataract family. *Ophthalmic Record* 5:489.

Hansell, H. P. 1900. A case of double retrobulbar optic neuritis, hereditary in origin. *Transactions of the American Ophthalmological Society* 9:114–116.

Harbitz, F. 1909. Multiple neurofibromatoses. *Archives of Internal Medicine* 3:32–65.

Harlan, H. 1898. A case of hereditary glaucoma. *Journal of the American Medical Association* 31:841–842.

Harrington, A. H. 1887–88. Hereditary cases of progressive muscular atrophy. *American Journal of Insanity* 44:73–76.

Harrington, H. L. 1885. A family record showing the heredity of disease. *Physician and Surgeon* 7:49–51.

Harris, S. N. 1847. Hereditary transmission. *Southern Journal of Medicine and Pharmacy* 2:516–527.

Hartshorne, H. 1885. General etiology, medical diagnosis, and prognosis. In *A System of Practical Medicine*, edited by W. Pepper, 1:125–172. Philadelphia: Lea.

Harvey, A. M. 1981. *Science at the Bedside: Clinical Research in American Medicine, 1905–1945*. Baltimore: Johns Hopkins University Press.

Hatch, F. F. 1915. Progressive neuromuscular atrophy (peroneal type of Charcot-Marie-Tooth). *Boston Medical and Surgical Journal* 172:393–398.

Hattie, W. H. 1909–10. Huntington's chorea. *American Journal of Insanity* 66:123–128.

Hay, C. M. 1889–90. Hereditary chorea. *University Medical Magazine* 2:463–472.

———. 1891. Hereditary chorea. *American Lancet* 15:284–287.

Hay, J. 1813. Account of a remarkable hemorrhagic disposition. *New England Journal of Medicine and Surgery* 2:221–225.

Hayes, D. 1841–43. Hereditary diseases. *Transactions of the Medical Society of New York* 5:97–107.

Hecht, D. 1913. The inheritance of epilepsy. *Medical Record* 84:323–329.

Henderson, C. R. 1909. Practical eugenics. *Proceedings of the American Breeders Association* 5:223–227.

Hereditary transmission and variation by descent. 1875. *Chicago Medical Journal and Examiner* 32:403–411.

Heredity of acquired characteristics. 1897. *Boston Medical and Surgical Journal* 137:427–428.

Herndon, C. N. 1956. Genetics in the medical school curriculum. *American Journal of Human Genetics* 8:1–7.

Heron, D. 1913. *Mendelism and the Problem of Mental Defect: A Criticism of Recent American Work*. London: University of London.

———. 1914. A rejoinder to Dr. Davenport. *Science* 39:24–25.

Herrman, C. 1915a. Amaurotic family idiocy in one of twins. *American Journal of Diseases of Children* 72:553–554.

———. 1915b. A case of amaurotic family idiocy in one of twins. *Archives of Pediatrics* 32:386–387.

———. 1915c. A case of amaurotic family idiocy in one of twins. *Archives of Pediatrics* 32:902–908.

———. 1917. The etiology of Mongolian imbecility. *Archives of Pediatrics* 34:494–503.

———. 1918. Heredity and disease. *Journal of Heredity* 9:77–80.

———. 1924. The relation of heredity to the diseases of infancy and childhood. *Archives of Pediatrics* 41:301–314.

Hertwig, O. 1884a. Das Problem der Befruchtung und der Isotropie des Eies, eine Theorie der Vererbung. *Zeitschrift fur Medizin und Naturwissenschaft* 18:21–23.

———. 1884b. Das Problem der Befruchtung und der Isotropie des Eies. *Zeitschrift fur Medizin und Naturwissenschaft* 18:276–318.

Hess, A. F. 1916. The blood and the blood vessels in hemophilia and other hemorrhagic diseases. *Archives of Internal Medicine* 17:203–220.

Hichney, F. 1912. Hermaphroditism, pseudo-hermaphroditism and differentiation of sex. *Medical Council* 17:112–115.

Hicks, H. 1903. Hemophilia. *Transactions of the Medical Association of Georgia*, 134–148.

Hilton, H. M. 1905. Heredity. *Medical Examiner and Practitioner* 15:737–743.

Hinchey, F. 1913. The principles of heredity with reference to disease. *Medical Fortnightly* 14:411–415.

Hirsch, W. 1898. The pathological anatomy of a fatal disease of infancy. *Journal of Nervous and Mental Diseases* 25:538–549.

His, W. 1874. *Unsere Korperform.* Leipzig: Vogel.

Hobshawn, E. 1987. *The Age of Empire, 1875–1914.* New York: Pantheon.

Hoffman, E. P., Fischbeck, K. H., and Brown, R. H. 1988. Dystrophin characterization in muscle biopsies from Duchenne and Becker muscular dystrophy patients. *New England Journal of Medicine* 318:1363–1368.

Hoke, J. 1889. On the philosophy of the transmission of hereditary disease. *North American Practitioner* 1:294–297.

Holden, C. 1991. Probing the complex genetics of alcoholism. *Science* 251:163–164.

Holloway, E. 1885. Hemophilia. *Cincinnati Lancet Clinic* 15:68–74.

Holmes, J. J., and Loomis, H. M. 1909–10. The heredity of eye color and hair color in man. *Biological Bulletin* 18:50–65.

Holt, L. E. 1887. Remarks upon spina bifida. *New York Medical Journal* 46:519–521.

Holtzapple, G. E. 1903–4. Family periodic paralysis. *Pennsylvania Medical Journal* 9:408–417.

Homan, G. 1898. Hereditary human cryptorchidism. *Medical Review* 38:444–446.

Hope, W. T. 1879–80. Congenital deformity, probably due to maternal impressions. *Virginia Medical Monthly* 6:882–883.

Hopkins, St. G. L. 1880–81. Hematophilia or the hemorrhagic diathesis. *Pacific Medical and Surgical Journal* 13:55–60.

Hoppe, H. H. 1903. Two cases of Friedreich's hereditary ataxia. *Cincinnati Lancet Clinic* 1:361–362.

Horne, B. S. 1902. The law of heredity. *Cincinnati Lancet Clinic* 48:86–87.

Howard, A. C. P. 1910. Amaurotic family idiocy. *Montreal Medical Journal* 39:429–431.

Howe, L. 1887. A family history of blindness from glaucoma. *Archives of Ophthalmology* 16:72–76.

———. 1911. Note on the heredity of corneal astigmatism. *Transactions of the American Ophthalmological Society* 12:1001–1004.

————. 1917. A note on the relation of sex linkage to certain hereditary diseases of the eye. *Transactions of the American Ophthalmological Society* 15:242–243.

————. 1918. The relation of hereditary eye diseases to genetics and eugenics. *Journal of the American Medical Association* 70:1994–1999.

————. 1919a. The relation of hereditary eye defects to genetics and eugenics. *Journal of Heredity* 10:379–382.

————. 1919b. Report of the Committee on Prevention of Hereditary Blindness. *Transactions of the Section on Ophthalmology of the American Medical Association*, 389.

————. 1925. Report of the Committee on Hereditary Blindness. *Transactions of the Section on Ophthalmology of the American Medical Association*, 363–367.

————. 1926. Concerning a law to lessen hereditary blindness. *Transactions of the American Ophthalmological Society* 24:106–111.

Hsu, T. C. 1979. *Human and Mammalian Cytogenetics*. New York: Springer.

Hubbard, O. S. 1914. Prevention of epilepsy. *Illinois Medical Journal* 26:366–368.

Huber, D. A., Buckler, A. J., and Glaser, T. 1990. An internal deletion within an 11p13 zinc finger gene contributes to the development of Wilm's tumor. *Cell* 61:1257–1269.

Huber, J. B. 1908. Heredity and disease. *Medical Times* 36:97–100.

Huddle, T. S. 1991. Looking backward: The 1871 reforms at Harvard Medical School reconsidered. *Bulletin of the History of Medicine* 65:340–365.

Hukk, D. L. 1988. *Science As a Process*. Chicago: University of Chicago Press.

Humphreys, W. C. 1968. *Anomalies and Scientific Theories*. San Francisco: Freeman, Cooper.

Humphries, P., Kenna, P., and Farrar, G. J. 1992. On the molecular genetics of retinitis pigmentosa. *Science* 256:804–808.

Hunt, E. H. 1910. Friedreich's ataxia. *American Journal of Obstetrics* 62:125–137.

Hunt, E. L. 1910. Friedreich's ataxia. *Archives of Pediatrics* 27:373.

————. 1911. Epilepsy. *Medical Record* 80:261–265.

Hunt, H. L. 1910. The heredity and congenital causes of exceptional development. *Medical Record* 78:48–52.

Huntington, G. 1872. On chorea. *Medical and Surgical Reporter* 26:317–321.

————. 1910. Recollections of Huntington's chorea as I saw it at East Hampton, Long Island, during my boyhood. *Journal of Nervous and Mental Diseases* 39:255–257.

Hutchins, M. B. 1893. The clinical history of a case of xeroderma pigmentosum. *Journal of Cutaneous and Genito-Urinary Diseases* 11:402–408.

Hutchinson, W. 1892a. Darwinism and disease. *Journal of the American Medical Association* 19:147–151.

————. 1892b. The influence of heredity in the prevention of disease. *Medical News* 60:169–173.

Hyde, J. N. 1903. Three cases of xeroderma pigmentosum. *Journal of Cutaneous Diseases* 21:573–575.

Hyer, R. N., Julier, C., and Buckley, J. D. 1991. High-resolution linkage mapping for susceptibility genes in human polygenic disease: Insulin-dependent diabetes mellitus and chromosome 11q. *American Journal of Human Genetics* 48:243–257.

Hymanson, A. 1902. A case of amaurotic family idiocy. *New York Medical Journal* 76:60–61.

———. 1913. Metabolism studies of amaurotic family idiocy. *Archives of Pediatrics* 30:825–836.

———. 1918. Diatheses in children. *New York Medical Journal* 107:872–874.

Iles, G. 1878. Heredity. *Popular Science Monthly* 11:356–364.

Ingham, S. D. 1915. Some considerations of heredity. *New York Medical Journal* 101:110–112.

The inheritance of acquired characters. 1919. *Journal of the American Medical Association* 73:838.

Irwell, L. 1902a. The nonheredity of acquired characteristics. *Medical News* 81:100–103.

———. 1902b. The nonheredity of acquired characters. *St. Louis Medical Review* 45:37–40.

———. 1912. Are acquired characters hereditary? *Medical Times* 40:189.

Is cancer either contagious or hereditary? 1917. *American Journal of Electrotherapeutics and Radiology* 35:423–425.

Iutzi, J. 1882. Heredity and its relation to disease. *Transactions of the Indiana Medical Society* 32:136–146.

Jacoby, G. W. 1887. Thomsen's disease. *Journal of Nervous and Mental Diseases* 14:129–151.

James, T. L. 1897. Heredity. *Dental Review* 11:145–151.

Jeffries, B. J. 1879. *Color Blindness*. Boston: Houghton.

Jenkins, D., Mijuvic, C., and Fletcher, J. 1990. Identification of susceptibility loci for type I diabetes. *Diabetologia* 33:387–395.

Jennings, H. S. 1924. Heredity and environment. *Scientific Monthly* 19:225–238.

Jennings, J. E. 1890. Four cases of albinism. *American Journal of Ophthalmology* 13:13–14.

Johnson, A. 1909. Race improvement by control of defectives (negative eugenics). *Annals of the American Academy of Political and Social Sciences* 34:22–29.

Johnson, C. 1905. Hereditary and maternal factors in health and disease. *Northwestern Lancet* 25:21–26.

Jones, D. F., and Mason, S. L. 1916. Inheritance of congenital cataract. *American Naturalist* 50:119–126, 751–757.

Jordan, D. S. 1907. Report of the Committee on Eugenics. *Proceedings of the American Breeders Association* 4:201–208.

———. 1909. Report of the Committee on Eugenics. *Proceedings of the American Breeders Association* 5:217–218.

———. 1914. Prenatal influences. *Journal of Heredity* 5:38–39.

Jordan, E. 1898. The inheritance of certain bacterial disease. *Chicago Medical Recorder* 15:82–86.

Jordan, E. 1992. The Human Genome Project: Where did it come from, where is it going? *American Journal of Human Genetics* 51:1–6.

Kagenoff, M. F. 1990. Understanding of the molecular basis of coeliac disease. *Gut* 31:497–499.

Kammerer, P. 1924. *The Inheritance of Acquired Characteristics*. New York: Boni and Liveright.

Kaufman, M. 1960. American Medical Education. In *The Education of American Physicians*, edited by R. L. Numbers, 12–15. Berkeley: University of California Press.

Kaufman, M., Galishoff, S., and Savitt, T. L. 1984. *Dictionary of American Medical Biography*. Westport, Conn.: Greenwood.

Kehoe, H. L. 1915. What is epilepsy? *Alienist and Neurologist* 36:7–11.

Keller, A. G. 1910. The limits of eugenics. *Bulletin of the American Academy of Medicine* 11:671–681.

Kellog, J. H. 1898. Early observations of the disease now known as Friedreich's ataxia. *Modern Medicine and Bacteriological Review* 7:273–274.

Kemp, J. 1809. Hereditary blindness. *Baltimore Medical and Physical Recorder* 1:273–277.

Keogh, C. H. 1916. Heredity: A study of an American genealogy. *Alienist and Neurologist* 37:369–374.

Kevles, D. J. 1980. Genetics in the U.S. and Great Britain, 1890–1930: A review with speculations. *Isis* 71:441–455.

———. 1985. *In the Name of Eugenics*. New York: Knopf.

Key, W. E. 1919. Better American families. *Journal of Heredity* 10:107–109.

Keyser, P. 1874. Ectopia lentis. *Medical and Surgical Reporter* 30:213.

Khomenko, A. G., Litvirov, V. I., and Chukamova, V. P. 1990. Tuberculosis in various HLA phenotypes. *Tubercule* 71:187–192.

Kiernan, J. W. 1902a. Heredity. *Medical News* 80:291–298.

———. 1902b. The problem of heredity. *Journal of the American Medical Association* 38:388–391.

Kimmelman, B. A. 1983. The American Breeders Association. *Social Studies in Science* 13:163–204.

King, B. 1991. Personal communication, Medical Library, Georgetown University School of Medicine, Washington, D.C.

King, C. 1885. Hereditary chorea. *New York Medical Journal* 41:468–470.

———. 1885–86. Another case of hereditary chorea. *Medical Press of Western New York* 1:674–677.

———. 1889. A third case of hereditary chorea. *Medical News* 55:39–41.

———. 1906. Hereditary chorea. *Medical Record* 70:765–768.

King, L. S. 1991. *Transformations in American Medicine*. Baltimore: Johns Hopkins University Press.

King, M. C. 1991. Localization of the early-onset breast cancer gene. *Hospital Practice* 26:121–126.

Kinzler, K. W., Nilbert, N. C., and Vogelstein, B. 1991. Identification of a gene located at chromosome 5q21 that is mutated in colorectal cancers. *Science* 251:1366–1370.

Kleeburg, V. R. 1989. Etiology and risk factors of melanoma. *Annals of Italian Chirurgery* 60:231–236.

Knapp, A. 1904. Hereditary optic atrophy. *Archives of Ophthalmology* 33:383–385.

Knapp, P. C. 1907. Heredity in diseases of the nervous system with especial reference to heredity in epilepsy. *Boston Medical and Surgical Journal* 157:485–490.

Koenig, M., Hoffman, E. P., and Bertelson, C. J. 1987. Complete cloning of the Duchenne muscular dystrophy (DMD) cDNA, and preliminary genomic organization of the DMD gene in normal and affected individuals. *Cell* 50:509–517.

Kohs, S. C. 1915. New light on eugenics. *Journal of Heredity* 6:446–452.

Krantz, D. L., and Wiggins, L. 1973. Personal and impersonal channels of recruitment in the growth theory. *Human Development* 16:133–156.

Krauss, W. C. 1902. Heredity with a study of statistics of the New York state hospitals. *American Journal of Insanity* 58:607–623.

Kroeber, A. L. 1916. The cause of the belief in use inheritance. *American Naturalist* 50:367–370.

Krumbhaar, E. B. 1930. John Conrad Otto and the recognition of hemophilia. *Johns Hopkins Hospital Bulletin* 46:123–140.

Kuh, S. 1900. A case of amaurotic family idiocy. *Journal of Nervous and Mental Diseases* 27:268–270.

Kuhn, T. S. 1970. *The Structure of Scientific Revolutions.* Chicago: University of Chicago Press.

Kunitz, S. J. 1983. The historical roots and ideological functions of disease concepts in three primary care specialties. *Bulletin of the History of Medicine* 57:412–432.

Laffer, W. B. 1909. Myatonia congenita of Oppenheim. *Ohio State Medical Journal* 5:609–616.

Lambert, J. A. 1899–1900. The influence of infectious and hereditary diseases upon the developing mind. *Denver Medical Times* 19:501–505.

Lanham, V. L. 1968. *Origins of Modern Biology.* New York: Columbia University Press.

Lanier, L. H. 1919–20. A study of congenital defects and heredity in relation to the eye. *Texas State Journal of Medicine* 15:424–425.

Larrabee, R. C. 1906. Hemophilia in the newly born. *American Journal of Medical Science* 131:497–505.

Laughlin, H. H. 1929. Letter to Mrs. L. Howe, November 4, 1929. Houghton Library, Harvard University, Cambridge, Mass.

Lawrence, G. A. 1910. Two cases of spastic paraplegia. *Post-Graduate* 25:66–68.

Layton, T. 1882. On the transmission and transformation of nervous diseases through heredity. *New Orleans Medical and Surgical Journal* 10:173–194.

LeConte, J. 1886. A case of inherited polydactylism. *Science* 8:166.

Levin, I. 1912. The influence of heredity on cancer. *Zeitschrift fur Krebsforschung* 11:547–550.

Lewis, A. C. 1915. A case of complete bilateral iridiremia. *Ophthalmic Record* 24:134–135.

Lewis, G. G. 1904. Hereditary ectopia lentis. *Archives of Ophthalmology* 33:275.

Lewis, J. B. 1860–63. Hereditary predisposition. *Proceedings of the Connecticut Medical Society* 1:87–106.

Libby, G. F. 1908. Consanguinity in relation to ocular disease. *Denver Medical Times* 28:112–117.

———. 1909. Hereditary ocular disease. *Denver Medical Times* 29:225–228.

Lightman, A., and Gingerich, O. 1991. When do anomalies begin? *Science* 255:690–696.

Lindgren, V., Bryke, C. R., and Ozcelik, T. 1992. Phenotypic, cytogenetic, and molecular studies of three patients with constitutional deletions of chromosome 5 in the region of the gene for familial adenomatous polyposis. *American Journal of Human Genetics* 50:988–997.

Little, C. C. 1915. Cancer and heredity. *Science* 42:218–219.

———. 1916. The relation of heredity to cancer in man and animals. *Scientific Monthly* 3:196–202.

———. 1922. The relation between research in human heredity and experimental genetics. *Scientific Monthly* 14:401–414.

Lloyd, J. H., and Newcomer, H. S. 1921. A case of Friedreich's ataxia. *Archives of Neurology and Psychiatry* 6:157–162.

Loeb, C. 1909. Hereditary blindness and its prevention. *Annals of Ophthalmology* 18:1–47, 245–314, 489–547, 755–789.

Longstreth, M. 1882. *Rheumatism, Joint, and Some Allied Disorders.* New York: Wood.

Lord, D., O'Farrell, A. G., and Staunton, H. 1990. The inheritance of multiple sclerosis susceptibility. *Irish Journal of Medical Science* 159:1–20.

Ludmerer, K. L. 1972. *Genetics and American Society: A Historical Appraisal.* Baltimore: Johns Hopkins University Press.

———. 1981. Reform at Harvard Medical School, 1869–1909. *Bulletin of the History of Medicine* 55:343–370.

———. 1985. *Learning to Heal: The Development of American Medical Education.* New York: Basic Books.

Lydston, J. A. 1890–91. Consanguinity in disease. *Denver Medical Times* 10:313–319.

Lyon, I. W. 1863. Chronic hereditary chorea. *American Medical Times* 7:289–290.

McClung, C. E. 1901. Notes on the accessory chromosome. *Anatomischer Anzeiger* 20:220–226.

McGaffin, C. G. 1911. A manic-depressive family: A study in heredity. *American Journal of Insanity* 68:263–269.

McGee, T. L., Yandell, T. P., and Dryja, T. P. 1990. Structure and partial genomic sequence of the human retinoblastoma susceptibility gene. *Gene* 80:119–128.

McHenry, L. C. 1969. *History of Neurology.* Springfield, Ill.: Thomas.

Mack, A. E. 1911. Heredity. *Western Medical Review* 16:449–455.

MacKay, H. J. 1894. The isolated type of Friedreich's disease. *American Journal of Medical Science* 108:151–158.

McKee, E. S. 1886. Consanguinity in marriage. *Medical Record* 30:4–7.

———. 1888. Cousin marriages unobjectionable. *Southern California Practitioner* 3:417–420.

McKee, J. H. 1905. A case of amaurotic family idiocy. *American Journal of Medical Science* 129:22–31.

MacKenzie, H. W. G. 1894. A case of non-hereditary Friedreich's disease. *American Journal of Medical Science* 107:371–374.

McKim, W. D. 1899. *Heredity and Human Progress.* New York: Putnam.

McKinnis, C. R. 1914. The value of eugenics in Huntington's chorea. *Medical Record* 86:103–106.

Macklin, M. T. 1933. The teaching of inheritance of disease. *Annals of Internal Medicine* 6:1336–1337.

McKusick, V. A. 1962. Hemophilia in early New England. *Journal of the History of Medicine* 17:342–365.

———. 1975. The growth of human genetics as a clinical discipline. *American Journal of Human Genetics* 27:261–273.

McMurrich, J. P. 1894. The phenomena of fertilization and their bearing on heredity. *Transactions of the Ohio Medical Society* 44:320–331.

McQuillen, J. H. 1870. Hereditary transmission of dental irregularities. *Dental Cosmos* 12:27–29, 73–75, 193–195.

Malkin, D., Li, F. P., and Strong, L. C. 1990. Germ line p53 mutations in a familial syndrome of breast cancer, sarcoma, and other neoplasms. *Science* 248:1233–1238.

Mall, F. P. 1987. In *William H. Welch and the Rise of Full-time Medicine*, edited by D. Fleming, 165. Baltimore: Johns Hopkins University Press.

Mandoff, G. T. 1904. Hereditary abnormality of the little finger. *New York Medical Journal* 80:642–643.

Mann, E. C. 1883. The nature and therapeutics of epilepsy. *Medical Bulletin* 5:4–8.

Margaritte, P., Bonati-Pellie, C., and King, M. C. 1992. Linkage of familial breast cancer to chromosome 17q21 may not be restricted to early-onset disease. *American Journal of Human Genetics* 50:1231–1234.

Martin, E. 1809. Hereditary blindness. *Baltimore Medical and Physical Recorder* 1:277–279.

Marvin, J. B. 1902–3. Muscular dystrophy. *Louisville Monthly Journal of Medicine and Surgery* 9:496–498.

Marx, J. 1989. Many gene changes found in cancer. *Science* 246:1386–1388.

———. 1990. Dissecting the complex diseases. *Science* 247:1540–1542.

———. 1993. New colon cancer gene discovered. *Science* 260:751–752.

Mason, V. R., and Rienhoff, W. F. 1920. Hereditary spastic paraplegia. *Johns Hopkins Hospital Bulletin* 31:215–217.

Maupertuis, P. L. M. de. 1768. *Oeuvres de Maupertuis.* Vol. 2. Lyon: Bruyset.

Maynard, H. H., and Scott, C. R. 1921. Hereditary multiple cartilaginous exostoses. *Journal of the American Medical Association* 76:579–581.

Mayr, E. 1982. *The Growth of Biologic Thought.* Cambridge, Mass.: Harvard University Press.

Meek, E. R. 1905. A case of neurofibroma. *Boston Medical and Surgical Journal* 152:370–372.

Mendel, G. 1865. Versuche uber Pflanzhybriden. *Verhandlungen des natur-forschender Verein in Brunn* 4:3–47.

Mendel's Law. 1907. *American Medicine* 2:132–135.

Merzbach, J. 1911. Remarkable history of family apoplexy. *Long Island Medical Journal* 5:166–167.

Meyers, E. L. 1904. Epilepsy: Its history, cause, and treatment. *Transactions of the Luzerne County Medical Society* 12:170–179.

Miles, H. S. 1896. A number of lens cases illustrating heredity. *Annals of Ophthalmology and Otology* 5:542–546.

Millikin, B. C. 1904. The hereditary element in cataract. *American Journal of Ophthalmology* 21:74–79.

Millikin, D. 1881. A bad family. *Cincinnati Lancet Clinic* 7:437–443.

Mills, C. K. 1898. Myotonia, or Thomsen's disease. In *An American Textbook of the Diseases of Children*, edited by L. Storr, 687–688. Philadelphia: Saunders.

Milroy, W. F. 1892–93. An undescribed variety of hereditary edema. *Omaha Clinic* 5:101–108.

Minot, C. S. 1886a. The physical basis of heredity. *Science* 8:125–130.

———. 1886b. Report on histology and embryology. *Boston Medical and Surgical Journal* 114:460–463.

Mitelman, F. 1991. Cancer cytogenetics: An overview. *American Journal of Human Genetics* 49:74.

Mix, C. L. 1903. Hereditary optic atrophy. *Chicago Medical Recorder* 24:175–181.

Moleen, G. A. 1919. Observation on heredity. *Colorado Medicine* 16:57–61.

Montgomery, C. M. 1916. Multiple cartilaginous exostoses. *International Clinics* 3:140–144.

Montgomery, T. H. 1904. The main facts in regard to the cellular basis of heredity. *Proceedings of the American Philosophical Society* 43:5–14.

———. 1910. Are particular chromosomes sex determinants? *Biological Bulletin* 19:1–17.

Moore, W. G. 1908. An atypical case of Friedreich's ataxia. *Journal of Nervous and Mental Diseases* 35:567–568.

Morgan, T. H. 1910. Chromosomes and heredity. *American Naturalist* 44:449–496.

———. 1924. Human inheritance. *American Naturalist* 58:385–409.

Morrison, W. H. 1881. Hemorrhagic diathesis. *Philadelphia Medical Times* 12:109–110.

Morrow, P. A. 1907. Hygiene in relation to the heredity of disease. *New York Medical Journal* 85:1162–1167.

Morse, H. C. 1985. The Bussey Institute and the early days of mammalian genetics. *Immunogenetics* 21:109–116.

Munson, J. F. 1910. An hereditary chart. *New York Medical Journal* 91:437.

———. 1917. Is epilepsy a bacterial infection? *New York Medical Journal* 105:836–837.

Murdoch, J. M. 1902–3. Notes on the family history of microcephalic children. *Journal of Psycho-Asthenics* 7:33–38.

Mussey, W. L. 1895. Ichthyosis. *Cincinnati Lancet Clinic* 34:657–659.

Nageli, C. 1884. *Mechanisch-physiologische Theorie der Abstammungslehre*. Leipzig: Oldenbourg.

Nammock, C. E. 1894. A case of non-hereditary Friedreich's disease. *Medical Record* 46:171–172.

Nasse, C. F. 1820. Von einer erblichen Neigung zu todlicher Blutungen. *Archiv für medizinishe Erfahrung im Gebiete der praktischen Medizin und Staatsarzneickunde*, 385–434.

Neate, J. S. 1909. The etiology and pathology of bilateral polycystic degenerate kidneys. *American Journal of Obstetrics* 60:61–77.

Neff, F. C. 1909. The diagnosis of Friedreich's disease. *Kansas City Medical-Index Lancet* 32:93–94.

Neff, I. H. 1894–95. A report of thirteen cases of ataxia in adults with hereditary history. *American Journal of Insanity* 51:365–373.

———. 1895–96. Some cases showing possible physical signs of degeneration. *American Journal of Insanity* 52:545–549.

———. 1905. A case of family ataxia of the hereditary cerebellar form with necroscopy. *Journal of the Michigan Medical Society* 4:328–332.

Neiberger, W. E. 1911. Human heredity. *Clinique* 32:692–699.

Nepom, G. T. 1990. A unified hypothesis for the complex genetics of HLA associations with IDDM. *Diabetes* 39:1153–1157.

Neu, C. F. 1911. Migraine. *Journal of the Indiana State Medical Association* 4:59–66.

Nickerson, C. L., Luthra, H. S., and David, C. S. 1990. Role of enterobacteria and HLA-B27 in spondyloarthropathies. *Annals of Rheumatologic Disease* 49:426–433.

Noble, C. P. 1909. Hereditary hypoplasia in man. *Journal of the American Medical Association* 52:552–555.

Norris, W. F. 1880–84. Hereditary atrophy of the optic nerves. *Transactions of the American Ophthalmological Society* 3:662–678.

———. 1882. Hereditary atrophy of the optic nerves. *Transactions of the American Ophthalmological Society*, 355–359.

———. 1886. Medical Ophthalmology. In *A System of Practical Medicine*, edited by W. Pepper, 4:737–804. Philadelphia: Lea.

Numbers, R. L. 1988. The fall and rise of the American medical profession. In *The Professions in American History*, edited by N. O. Hatch, 51–72. South Bend, Ind.: University of Notre Dame Press.

Nunn, R. 1895. Are acquired characters transmitted to offspring? *Medical Sentinel* 3:421–432.

Ogier, T. L. 1848. Hereditary predisposition. *Charleston Medical Journal* 3:261–267.

Oksenberg, J. R., and Steinman, L. 1989–90. The role of the MHC and T-cell receptor in susceptibility to multiple sclerosis. *Current Opinion in Immunology* 2:619–621.

Olby, R. 1985. *Origins of Mendelism.* Chicago: University of Chicago Press.

Osborn, H. F. 1889. The paleontological evidence for the transmission of acquired characters. *American Naturalist* 23:561–566.

———. 1892. Present problems in evolution and heredity. *Medical Record* 41:197–204, 253–260, 449–456, 533.

Osler, W. 1880. On heredity in progressive muscular atrophy as illustrated by the Farr family of Vermont. *Archives of Medicine* 4:316–320.

———. 1885. Hemophilia. In *A System of Practical Medicine*, edited by W. Pepper, 3:931–939. Philadelphia: Lea.

———. 1892. *The Principles and Practice of Medicine.* New York: Appleton.

———. 1893. Remarks on the varieties of chronic chorea, and a report upon two families of the hereditary form with one autopsy. *Journal of Nervous and Mental Diseases* 18:97–111.

———. 1894a. Case of hereditary chorea. *Johns Hopkins Hospital Bulletin* 5:119–120.

———. 1894b. *On Chorea and Choreiform Affections.* Philadelphia: Blakiston.

———. 1907. *Modern Medicine.* Vol. 1. Philadelphia: Lea and Febiger.

Osler, W., and McCrae, T. 1913. *Modern Medicine: Its Theory and Practice.* Vol. 1. Philadelphia: Lea and Febiger.

Otto, J. C. 1803. An account of an hemorrhagic disposition existing in certain families. *Medical Repository* 6:1–4.

Ounton, J. 1909. Hereditary spastic paraplegia: Report of seven cases in two families. *Kansas City Medical-Index Lancet* 32:371–375, 407–411.

Outter, W. B. 1870. On the relationship existing between some forms of hereditary disease. *St. Louis Medical and Surgical Journal* 7:54–60.

Packard, F. R. 1931. *History of Medicine in the United States*. New York: Hoeber.

Painter, T. S. 1921. The Y chromosome in mammals. *Science* 53:503–504.

———. 1922. The spermatogenesis of man. *Anatomic Record* 23:129.

———. 1923. Studies in mammalian spermatogenesis. II. The spermatogenesis of man. *Journal of Experimental Zoology* 37:291–336.

Pallen, J. S. 1856. Heritage, an hereditary transmission. *St. Louis Medical and Surgical Journal* 14:490–501.

Parish, L. C. 1976. The status of the history of dermatology: An American appraisal. *International Journal of Dermatology* 15:351–354.

Parker, E. H. 1877. Heredity as a factor in pauperism and crime. *Transactions of the Medical Society of New York*, 158–171.

Parker, E. S. 1898. Five cases of congenital bilateral symmetric displacement of the lens of the eye in three successive generations in one family. *Journal of the American Medical Association* 31:708–710.

Parker, W. R. 1916a. A history of family cataracts through four generations. *Journal of the Michigan Medical Society* 15:188–190.

———. 1916b. A history of family cataracts through four generations. *Transactions of the Clinical Society of the University of Michigan* 7:51–53.

Paul, D. B., and Kimmelman, B. A. 1988. Mendel in America: Theory and practice, 1900–1919. In *The American Development of Biology*, edited by R. Rainger, K. R. Benson, and J. Maienschein, 281–312. Philadelphia: University of Pennsylvania Press.

Pauly, P. 1984. The appearance of academic biology in late nineteenth-century America. *Journal of the History of Biology* 17:369–397.

Peltomaki, P., Aaltonen, L. A., and Sistonen, P. 1993. Genetic mapping of a locus predisposing to human colorectal cancer. *Science* 260:810–812.

Pennington, J. 1877. Hereditary transmission of disease. *Transactions of the Indiana State Medical Society* 27:113–118.

Percy, N. M. 1915. Multiple chondro-osteomas. *Surgery Gynecology Obstetrics* 20:619–621.

Perdue, E. M. 1914. A new contribution to the etiology and pathogenesis of cancer. *Medical Council* 19:258–263.

Perry, M. L. 1911–12. Prevention of epilepsy. *Journal of the Missouri Medical Association* 8:379–381.

Peters, S. 1879. Family traits. *Medical and Surgical Reporter* 41:184–188.

Petersen, D. D., McKinney, C. E., and Ikeya, K. 1991. Human CYP1A1 gene: Cosegregation of the enzyme inducibility phenotype and an RFLP. *American Journal of Human Genetics* 48:720–725.

Peterson, M. J. 1978. *The Medical Profession in Mid-Victorian London*. Berkeley: University of California Press.

Phelps, R. M. 1892. A new consideration of hereditary chorea. *Journal of Nervous and Mental Diseases* 19:765–776.

Phillips, J. 1908. Cephalocoele with a report of three cases in one family. *Medical Record* 74:617–620.

Phillips, J. M. 1922. Angioneurotic edema. *Journal of the American Medical Association* 78:497–499.

Pickering, G. W. 1976. The view from the United Kingdom. In *Advances in American Medicine: Essays at the Bicentennial*, edited by J. Z. Bowers and E. F. Purcell, 774–788. New York: Macy Foundation.

Poore, C. T. 1875. Pseudohypertrophic muscular paralysis. *New York Medical Journal* 21:569–596.

Porter, J. L. 1897. Two cases of pseudohypertrophic paralysis. *Chicago Medical Recorder* 12:416–418.

Posey, W. C. 1898. Hereditary optic nerve atrophy. *Annals of Ophthalmology* 7:357–371.

Potter, J. D. 1992. Reconciling the epidemiology, physiology, and molecular biology of colon cancer. *Journal of the American Medical Association* 268:1573–1577.

Powers, G. H. 1892. Congenital cataract. *Pacific Medical Journal* 42:399–400.

Preiser, S. A., and Davenport, C. B. 1918. Multiple neurofibromatoses. *American Journal of Medical Science* 156:507–540.

Prentiss, C. W. 1902–3. Polydactylism in man and domestic animals. *Bulletin of the Museum of Comparative Zoology* 40:243–314.

Pringsheim, N. 1856. Observations sur le fecondation et la generation alternate algues. *Annales Science et Botanique* 5:256–258.

Prudden, L. E. 1927. *Biographical Sketches and Letters of T. Mitchell Prudden, M.D.* New Haven: Yale University Press.

Prudden, T. M. 1882. Cell life and animal life. *Medical News* 41:421–424.

Punnett, R. C. 1911. *Mendelism*. London: Macmillan.

———. 1912. Genetics and eugenics. *Problems in Eugenics. Papers Communicated to the First International Eugenics Congress*, 137.

Pusey, W. A. 1933. *The History of Dermatology*. Springfield, Ill.: Thomas.

Pusey, W. B. 1892. Unusual family history in two cases of glaucoma. *Archives of Ophthalmology* 21:360–361.

Rachford, B. K. 1892. Heredity. *Medical News* 60:227–228.

Ransohoff, J. 1913. Heredity in bone lesions. *Transactions of the Southern Surgical and Gynecologic Association* 26:166–170.

Ravin, A. 1985. Genetics in America: An historical overview. In *Genetic Perspectives in Biology and Medicine*, edited by E. D. Garber, 17–34. Chicago: University of Chicago Press.

Ravogli, A. 1917. Considerations on epidermolysis bullosa. *Journal of the American Medical Association* 69:256–260.

Ray, B. L. 1862. Hereditary transmission. *Communications of the Rhode Island Medical Society* 1:139–157.

Réaumur, R. A. F. de. 1751. *L'art de faire éclore et d'enlever en toute saison des oiseaux domestiques des toutes espèces*, Paris: Impr. Royale.

Reber, W. 1895. Six instances of color-blind women occurring in two generations of one family. *Medical News* 66:95–97.

Redfield, C. L. 1916a. Further remarks on the inheritance of acquired characteristics. *Cincinnati Lancet Clinic* 116:97–100.

———. 1916b. Method of testing of inheritance of acquirements. *Cincinnati Lancet Clinic* 116:409–412.

———. 1917. The inheritance of acquirements. *Maryland Medical Journal* 55:159–163.

Reed, S. C. 1979. A short history of human genetics in the U.S.A. *American Journal of Medical Genetics* 3:282–295.

Reuben, M. S. 1914. Splenohepatomegaly of Gaucher. *American Journal of Diseases of Children* 8:336–356.

———. 1918. Gaucher's disease. *New York Medical Journal* 107:118–120.

Reynolds, A. R. 1896. Degeneracy: Its causes and prevention. *Journal of the American Medical Association* 27:953–955.

Ribot, T. 1875. *Heredity*. New York: Appleton.

Richardson, A. B. 1890–91. The transmission of acquired variations. *American Journal of Insanity* 47:397–409.

———. 1896. The influence of heredity. *Transactions of the Ohio Medical Society*, 118–129.

Riggs, C. E. 1896. Two cases of Friedreich's ataxia. *Northwestern Lancet* 16:264–265.

Riley, W. H. 1902. The influence of heredity in nervous diseases. *Good Health* 37:218–221.

Riordan, J. R., Rommens, J. M., and Kerem, B. 1989. Identification of the cystic fibrosis gene: Cloning and characterization of the complementary DNA. *Science* 245:1066–1073.

Risley, S. D. 1915a. Aniridia with interesting family history. *Transactions of the College of Physicians of Philadelphia* 37:397–400.

———. 1915b. Hereditary aniridia. *Journal of the American Medical Association* 64:1310–1312.

Robin, E. A. 1909–10. The occurrence of several cases of juvenile cataract in one family. *New Orleans Medical and Surgical Journal* 62:931–933.

Robins, W. L. 1906–7. Friedreich's disease. *Washington Medical Annals* 5:378–390.

Rogers, A. C. 1909–10. Modern studies in heredity. *Journal of Psycho-Asthenics* 14:117–121.

Rogers, S. 1869. Hereditary diseases of the nervous system, unattended by mental aberration. *Quarterly Journal of Psychological Medicine* 3:625–647.

Rommens, J. M., Iannuzzi, M. C., and Kerem, B. 1989. Identification of the cystic fibrosis gene: Chromosome walking and jumping. *Science* 245:1059–1065.

Rosanoff, A. J. 1912. Inheritance of the neuropathic constitution. *Journal of the American Medical Association* 58:1266–1269.

Rosanoff, A. J., and Martin, H. E. 1915. Offspring of the insane. *Journal of Heredity* 6:355–356.

Rosanoff, A. J., and Orr, F. I. 1911a. *A Study of Heredity of Insanity in the Light of Mendelian Theory*. Cold Spring Harbor, N.Y.: Eugenics Record Office.

———. 1911b. A study of heredity of insanity in the light of the Mendelian theory. *American Journal of Insanity* 68:221–261.

Rosenberg, C. E. 1961. Charles B. Davenport and the beginning of human genetics. *Bulletin of the History of Medicine* 35:266–276.

————. 1967. Factors in the development of genetics in the United States. *Journal of the History of Medicine* 22:27–46.

————. 1974. The bitter fruit: Heredity, disease, and social thought in nineteenth-century America. *Perspectives in American History* 8:189–235.

Rothstein, W. G. 1972. *American Physicians in the 19th Century*. Baltimore: Johns Hopkins University Press.

————. 1987. *American Medical Schools and the Practice of Medicine*. New York: Oxford University Press.

Roux, W. 1883. *Über die Bedeutung der Kerntheilungsfiguren*. Leipzig: Engelmann.

Rowland, A. A. 1881. Pseudohypertrophic paralysis. *St. Louis Medical and Surgical Journal* 40:583–588.

Ruchelli, E. D., Horn, M., and Taylor, S. R. 1990. Severe chemotherapy-related hepatic toxicity associated with MZ protease inhibitor phenotype. *American Journal of Pediatric Hematology and Oncology* 12:351–354.

Rumbald, T. F. 1894. Disease is not inherited. *Transactions of the Medical Society of California*, 52–60.

Russell, E. S. 1954. One man's influence: A tribute to William E. Castle. *Journal of Heredity* 45:210–213.

Russell, N. 1986. *Like Engendering Like: Heredity and Animal Breeding in Early Modern England*. Cambridge: Cambridge University Press.

Sachs, B. 1887. On arrested cerebral development with special reference to its cortical pathology. *Journal of Nervous and Mental Diseases* 14:541–553.

————. 1903. On amaurotic family idiocy. *Journal of Nervous and Mental Diseases* 30:1–13.

————. 1915. Amaurotic family idiocy. In *Modern Medicine*, edited by W. Osler and T. McCrae, 5:863–870. Philadelphia: Lea and Febiger.

Sachs, B., and Strauss, I. 1910. The cell changes in amaurotic family idiocy. *Journal of Experimental Medicine* 12:685–695.

Sager, J. 1900. Heredity. *Cincinnati Lancet Clinic* 44:198–200.

St. George-Hyslop, P. H. 1990. Genetic linkage studies suggest that Alzheimer's disease is not a single homogeneous disorder. *Nature* 347:194–197.

Sanborn, F. B. 1898. *Memoirs of Pliny Earle, M.D.* Boston: Damrell and Upham.

Sapp, J. 1983. The struggle for authority in the field of heredity. *Journal of the History of Biology* 16:311–342.

Scheinfeld, A. 1940. Genetics and the medical man. *Human Heredity* 5:129.

Schneck, J. 1879. Hereditary variation in the radial arteries. *Chicago Medical Journal and Examiner* 39:475–476.

Schwartz, D. 1909. Ichthyosis. *Journal of Cutaneous Diseases* 27:120–121.

Sedgwick, J. P. 1910. Von Graefe's sign in myotonia congenita. *American Journal of Medical Science* 140:80–86.

Sentz, L. 1991. Personal communication, History of Medicine Collection, Health Sciences Library, State University of New York at Buffalo.

Servos, J. W. 1986. Mathematics and the physical sciences in America, 1880–1930. *Isis* 77:611–629.

Shannon, G. A. 1913. Notes on Huntington's chorea. *Canadian Medical Association Journal* 3:962–964.

Shastid, T. S. 1917. History of ophthalmology. In *The American Encyclopedia and*

Dictionary of Ophthalmology, edited by C. A. Wood, 8775–8904. Chicago: Cleveland Press.

Shattuck, G. B. 1888. Three cases of "hereditary" locomotor ataxia (Friedreich's disease). *Boston Medical and Surgical Journal* 118:168–169.

Sheets, A. 1889. Heredity. *Proceedings of the Connecticut Medical Society,* 195.

Sherwood, F. R. 1899. Congenital deformities of the limbs. *Journal of the American Medical Association* 32:515–520.

Shike, M., Winower, S. J., and Greenwald, P. H. 1990. Primary prevention of colorectal cancer. *Bulletin of the World Health Organization* 68:377–385.

Shoemaker, J. V. 1907. Ichthyosis. *Medical Bulletin* 29:365–366.

Shryock, R. H. 1953. *The Unique Influence of the Johns Hopkins University on American Medicine.* Copenhagen: Munksgaard.

Shute, D. K. 1897. Heredity with variation. *New York Medical Journal* 66:341–346.

Sinkler, W. 1885. Two cases of Friedreich's disease. *Medical News* 47:3–6.

———. 1889. Two additional cases of hereditary chorea. *Journal of Nervous and Mental Diseases* 14:69–91.

———. 1906. Friedreich's ataxia. *New York Medical Journal* 83:65–72.

———. 1907. A case of epilepsy of the family type. *New York Medical Journal* 86:1067.

Sinkler, W., and Pearce, F. S. 1900. Family diseases. *Journal of the American Medical Association* 34:331–335.

Sisson, E. O. 1894. Heredity and evolution as they should be viewed by the medical profession. *Medical Review* 30:304–308.

Slye, M. 1915. The influence of heredity upon the occurrence of spontaneous cancer. *Interstate Medical Journal* 22:692–721.

———. 1916. Cancer and heredity. *Science* 43:135–136.

Small, S. I. 1895. Friedreich's ataxia. *Medical Record* 48:85–87.

Smith, A. L. 1908. The explosion of the theory of heredity. *New York Medical Journal* 88:529–533.

Smith, E. H. 1805. Account of a singular case of hemorrhage. *Philadelphia Medical Museum* 1:284.

Smith, H. 1898. History of a case of Huntington's chorea. *Medical Record* 54:422–423.

Smith, R. M. 1912. Amaurotic family idiocy. *Boston Medical and Surgical Journal* 166:370–372.

Smith, W. E. 1885. Hereditary or degenerative ataxia. *Boston Medical and Surgical Journal* 113:361–368.

———. 1888. Postero-lateral spinal sclerosis. *Boston Medical and Surgical Journal* 118:213–215.

Snediger, W. S. 1896. Heredity. *Pacific Medical Journal* 39:69–75.

Snyder, L. H. 1951. Old and new pathways in human genetics. In *Genetics in the 20th Century,* edited by L. C. Dunn, 369–391. New York: Macmillan.

Southworth, C. N. 1908. Heredity in disease. *New York State Medical Journal* 8:350–353.

Spencer, H. 1864. *Principles of Biology.* London: Williams and Norgate.

———. 1893. The inadequacy of natural selection. *Contemporary Review* 63:152–166, 439–456.

Spiece, W. K. 1919. A case of ectopia lentis. *Illinois Medical Journal* 35:1000–1001.

Spiller, W. G. 1902. Fourteen cases of spastic spinal paralysis occurring in one family. *Philadelphia Medical Journal* 9:1129–1131.

———. 1910a. Friedreich's ataxia. *Journal of Nervous and Mental Diseases* 37:411–435.

———. 1910b. Hereditary spastic spinal paralysis. In *Modern Medicine*, edited by W. Osler, 7:106–109. Philadelphia: Lea.

———. 1910c. Progressive muscular dystrophy. In *Modern Medicine*, edited by W. Osler, 7:119–125. Philadelphia: Lea.

———. 1915a. Friedreich's ataxia. In *Modern Medicine*, edited by W. Osler and T. McCrae, 5:130–139. Philadelphia: Lea and Febiger.

———. 1915b. Hereditary spastic spinal paralysis. In *Modern Medicine*, edited by W. Osler and T. McCrae, 5:105–108. Philadelphia: Lea and Febiger.

———. 1915c. Progressive neural muscle atrophy Charcot-Marie-Tooth type. In *Modern Medicine*, edited by W. Osler and T. McCrae, 5:109–112. Philadelphia: Lea and Febiger.

———. 1916. The family form of pseudo-sclerosis and other conditions attributed to the lenticular nucleus. *Journal of Nervous and Mental Diseases* 43:23–36.

Spillman, W. J. 1905. Mendel's law in relation to animal breeding. *Proceedings of the American Breeders Association* 1:171–177.

Spiro, L., Otterud, B., and Stauffer, D. 1992. Linkage of a variant or attenuated form of adenomatous polyposis coli to the adenomatous polyposis coli (APC) locus. *American Journal of Human Genetics* 51:92–100.

Spitzka, E. C. 1885. The family form of tabes dorsalis. In *A System of Practical Medicine*, edited by W. Pepper, 5:870–873. Philadelphia: Lea.

Spratling, W. P. 1894. On epilepsy in early life. *Medical News* 65:291–295.

———. 1910. Epilepsy. In *Modern Medicine*, edited by W. Osler, 7:660–661. Philadelphia: Lea.

Stanton, F. L. 1914. Heredity. *Dental Cosmos* 56:1049–1055.

Starr, M. A. 1898. Friedreich's ataxia. *Journal of Nervous and Mental Diseases* 25:194–199.

Starr, P. 1982. *The Social Transformation of American Medicine*. New York: Harper.

Stearns, H. P. 1897. Insane heredity. *Alienist and Neurologist* 18:1–9.

Steele, A. J. 1879. Pseudohypertrophy musculorum. *Transactions of the Medical Association of Missouri* 22:100–105.

Steiner, W. R. 1900. Hemophilia in the Negro. *Johns Hopkins Hospital Bulletin* 11:44–47.

———. 1915. Myotonia congenita: Thomsen's disease. In *Modern Medicine*, edited by W. Osler and T. McCrae, 5:886–893. Philadelphia: Lea and Febiger.

———. 1917. Hereditary hemorrhagic telangiectasiae. *Archives of Internal Medicine* 19:194–219.

Stephens, L. C. 1892. Hereditary chorea. *Transactions of the South Carolina Medical Association* 101–106.

Stevens, N. M. 1905–6. *Studies in spermatogenesis with especial reference to the accessory chromosome*. Publication no. 36. Washington, D.C.: Carnegie Institution.

Stockton-Hough, J. 1875. The laws of transmission of resemblance from parents to their children. *Medical Record* 8:409–414, 465–468, 521–526, 577–598.

Stoller, J. H. 1890. Human heredity. *Popular Science Monthly* 37:359–365.

Strauss, A. 1910. Amaurotic family idiocy. *American Journal of Obstetrics* 61:340–342.

Strong, T. J. 1906. A case of hypospadius through five generations. *Vermont Medical Monthly* 12:125–126.

Stubbe, H. 1972. *History of Genetics*. Cambridge, Mass.: MIT Press.

Sturtevant, A. H. 1965. *A History of Genetics*. New York: Harper and Row.

Su, L. K., Kinzler, K. W., and Vogelstein, B. 1992. Multiple endocrine neoplasia caused by a mutation in the murine homolog of the APC gene. *Science* 256:668–670.

Surber, D. 1900. Heredity and parental influence. *Chicago Medical Times* 33:422–425.

Sutter, W. P. 1991. Book review of *Genome*, by J. E. Bishop and W. Waldholz. *Pharos* 54:45.

Sutton, W. S. 1903. The chromosomes in heredity. *Biological Bulletin* 4:231–251.

Sweeney, A. 1912. Retinitis pigmentosa. *St. Paul Medical Journal* 14:219–221.

Tait, B. D. 1990. Genetic susceptibility in type 1 diabetes. *Journal of Autoimmunity* 3:3–11.

Talbot, E. S. 1898. Heredity and atavism. *Alienist and Neurologist* 19:628–658.

———. 1904. Etiology of cleft palate. *Medicine* 10:742–748.

Taylor, E. W. 1898. Family periodic paralysis. *Journal of Nervous and Mental Diseases* 25:637–660, 719–746.

Taylor, R. W. 1888. Xeroderma pigmentosum. *Medical News* 52:162–164.

Temkin, O. 1971. *The Falling Sickness: A History of Epilepsy*. Baltimore: Johns Hopkins University Press.

Terry, W. 1918a. Hereditary osteopsathyrosis. *Annals of Surgery* 68:231–234.

———. 1918b. Hereditary osteopsathyrosis. *Transactions of the American Surgical Association* 36:317–325.

Thibideau, S. N., Bren, G., and Schmid, D. 1993. Microsatellite instability in cancer of the proximal colon. *Science* 260:816–819.

Thom, D. A. 1915. The relation between the genetic factors and the age of onset in 157 cases of hereditary epilepsy. *Boston Medical and Surgical Journal* 173:469–473.

Thompson, J. L. 1887. Observations on displacement of the crystalline lens from congenital and other causes. *Journal of the American Medical Association* 9:674–676.

Thomson, J. A. 1908. *Heredity*. New York: Putnam.

Tiffany, F. B. 1895a. Ectopia lentis. *Journal of the American Medical Association* 25:796–798.

———. 1895b. Ectopia lentis. *Ophthalmic Record* 5:129–134.

Tilney, F. 1908. A family in which the choreic strain may be traced back to colonial Connecticut. *Neurographs* 1:124–127.

Timme, A. R. 1914. The hereditary basis of eugenics. *Cleveland Medical Journal* 15:595–606.

Timme, W. 1917. Progressive muscular dystrophy as an endocrine disease. *Archives of Internal Medicine* 19:79–104.

Tomlinson, H. A. 1879–80. The existence of supernumary fingers running through three generations. *Medical Times* 10:212–213.

Toomey, N. 1916. A family with myotonia, probably intermittent form of Thomsen's disease. *American Journal of Medical Science* 152:738–746.

Tracy, S. W. 1992. George Draper and American constitutional medicine, 1916–1946: Re-inventing the sick man. *Bulletin of the History of Medicine* 66:53–87.

Van Beneden, E. 1883. Recherches sur la maturation de l'oeuf et la féconda-tion. *Archives de Biologie* 4:265.

Van der Hoeve, J. 1909. Coloboma at the nerve head with normal vision. *Archives of Ophthalmology* 38:147–152.

Van Wagener, B. 1914. Surgical sterilization as a eugenic measure. *Journal of Psycho-Asthenics* 18:185–188.

Van Wort, R. M. 1904. Friedreich's ataxia. *New York Medical Journal* 80:1263–1264.

Von Schaick, G. G. 1888. The nature of inherited disease. *New York Medical Journal* 47:141–142.

Von Tschermak, E. 1900. Über kunstliche Kreuzung bei Pisum sativum. *Berichte der Deutscher botanischer Gesellschaft* 18:232–239.

Waldeyer, W. 1888. Über Karyokinese und ihre Beziehungen zu den Befruchtungsvorgangen. *Archiv fur mikroscopishe Anatomie* 32:48.

Walker, D. H. 1910. Ménière's disease: Report of a series of cases occurring in one family. *Transactions of the American Otologic Society* 12:150–157.

Walker, G. 1901. Remarkable case of hereditary ankyloses. *Johns Hopkins Hospital Bulletin* 12:129–132.

Walker, R. A. 1897–98. Heredity. *Pennsylvania Medical Journal* 1:409–415.

Walsh, W. S. 1918. Maternal impressions and feeble-mindedness. *Medical Record* 94:1113–1116.

Wandless, H. W. 1909. Amaurotic family idiocy. *New York Medical Journal* 89:953–954.

Ward, C. W. 1907. Report of Committee on Theoretical Research in Heredity. *Proceedings of the American Breeders Association* 3:130–132.

Warner, J. H. 1985. The selective transport of medical knowledge: Antebellum American physicians and Parisian medical therapeutics. *Bulletin of the History of Medicine* 59:213–231.

———. 1986. *The Therapeutic Perspective.* Cambridge, Mass.: Harvard University Press.

———. 1991. Ideals of science and their discontentments in late nineteenth-century American medicine. *Isis* 82:454–478.

Warthin, A. S. 1913. Heredity with reference to carcinoma. *Archives of Internal Medicine* 12:546–555.

———. 1914. Family susceptibility to cancer. *Transactions of the Clinical Society of the University of Michigan* 5:21–26.

Waterman, P. 1920. Heredity. *Proceedings of the Connecticut Medical Society* 128:152–178.

Waters, C. O. 1842. Chorea. In *The Practice of Medicine*, edited by R. Dunglison, 2:312. Philadelphia: Blanchard.

Watson, J. D. 1990. The Human Genome Project: Past, present, and future. *Science* 248:44–49.

Watson, J. D., and Cook-Deegan, R. M. 1990. The Human Genome Project and international health. *Journal of the American Medical Association* 263:3322–3324.

Weber, P. J. 1906. A case of Friedreich's ataxia. *Medical Fortnightly* 30:650–652.

Weber, W. W. 1990. Acetylation. *Birth Defects* 26:43–65.

Webster, D. 1878. The etiology of retinitis pigmentosa. *Transactions of the American Ophthalmological Society* 2:495–505.

Weeks, D. F. 1912. The heredity of epilepsy analyzed by the Mendelian method. *Proceedings of the American Philosophical Society* 51:178–190.

———. 1915. Epilepsy with special reference to heredity. *Journal of the Medical Society of New Jersey* 12:542–549.

Weeks, J. E. 1903. The influence of heredity on the eye. *Medical Record* 64:205–207.

Weinberg, R. A. 1991. Tumor suppressor genes. *Science* 254:1138–1146.

Weismann, A. 1891–92. On Heredity. *Essays upon Heredity*, edited by E. B. Poulton, 1:81–82. Oxford: Clarendon.

———. 1892. *Das Keimplasma: Eine Theorie der Vererbung*. Jena: Fischer.

———. 1913. *Vortrage über Deszendenztheorie*. Vol. 2. Jena: Fischer.

Welch, W. H. 1885. Cancer of the stomach. In *A System of Practical Medicine*, edited by W. Pepper, 2:530–579. Philadelphia: Lea.

Wells, E. A. 1888a. Friedreich's disease: Report of a group of five cases. *Journal of the American Medical Association* 11:303–304.

———. 1888b. Report of a group of five cases of Friedreich's disease. *Transactions of the Indiana Medical Society* 39:152a–d.

Welt-Kakels, S. 1917. Amaurotic family idiocy. *American Journal of Obstetrics* 76:1032–1033.

Whetstone, M. S. 1902. Heredity. *Dietetics and Hygienic Gazette* 18:129–132.

White, C. J. 1896. Dystrophia unguium et pilorum hereditaria. *Journal of Cutaneous and Genito-Urinary Diseases* 14:220–227.

White, W. A. 1909. Some thoughts on heredity. *Long Island Medical Journal* 3:237–244.

———. 1911. Letter to C. B. Davenport, May 22, 1911. Box 6, White Papers, Record Group 418, National Archives, Washington, D.C.

———. 1913. Eugenics and heredity in nervous and mental disorders. In *Modern Treatment of Nervous and Mental Disease*, edited by W. A. White, 1:17–55. Philadelphia: Lea and Febiger.

———. 1916. Letter to A. P. Herring, November 4. In G. N. Grog, *The Inner World of Psychiatry*, 171–172. New Brunswick, N.J.: Rutgers University Press.

———. 1919. Letter to A. Fisher, November 19. In G. N. Grog, *The Inner World of Psychiatry*, 172–173. New Brunswick, N.J.: Rutgers University Press.

———. 1938. *William Alanson White: The Autobiography of a Purpose*. New York: Doubleday.

Whitehouse, W. H. 1906. Three cases of prurigo of Hebra in one family. *Journal of Cutaneous Diseases* 24:177–178.

Whittaker, J. T. 1880. Haemorrophilia. *Cincinnati Lancet Clinic* 5:263–266.

Why the babies die. 1918. *Journal of Heredity* 9:62–66.

Wieman, H. L. 1912. Chromosomes in man. *American Journal of Anatomy* 14:461–471.

———. 1917. The chromosomes of human spermatocytes. *American Journal of Anatomy* 21:1–21.

Wilcox, E. V. 1900. Human spermatogenesis. *Anatomischer Anzeiger* 17:316–318.

———. 1901. Longitudinal and transverse division of chromosomes. *Anatomischer Anzeiger* 19:332–335.

Wilkie, J. S. 1962. Some reasons for the rediscovery and appreciation of Mendel's work in the first years of the present century. *British Journal of the History of Science* 1:5–17.

William, S. 1907. Keratosis palmae hereditaria. *Medical Record* 71:416.

Williams, A. D. 1880. A remarkable descent of congenital cataract. *St. Louis Medical and Surgical Journal* 38:368–369.

Williams, R. R., Hunt, S. E., and Husstedt, S. J. 1989. Current knowledge regarding the genetics of human hypertension. *Journal of Hypertension* 7:8–13.

Williams, T. A. 1915. Spinal gliosis. *Journal of Nervous and Mental Diseases* 40:186.

Willier, B. H. 1974. Charles H. Danforth. *Biographical Memoirs of the National Academy of Sciences* 44:1–56.

Wilmanth, A. N. 1910–11. Results of heredity and their bearing on poverty, crime, and disease. *Wisconsin Medical Journal* 9:257–269.

Wilson, E. B. 1902. Mendel's principles of heredity and the maturation of germ cells. *Science* 16:991–993.

———. 1907a. Recent studies in heredity. *Journal of the American Medical Association* 48:1557–1563.

———. 1907b. Recent studies in heredity. *Medical Record* 71:192–193.

———. 1908. Some recent studies in heredity. In *The Harvey Lectures, 1906–1907*, 200–221.

Wilson, H. 1891. Hereditary congenital cataract. *Journal of Ophthalmology, Otology, and Laryngology* 3:291–294.

Withington, C. F. 1885. Consanguinous marriages: Their effect upon offspring. *Boston Medical and Surgical Journal* 113:172.

Witter, G. F. 1887. Heredity. *Wisconsin State Board of Health Report* 10:167–181.

Wood, C. A. 1892. A picture of hereditary nystagmus. *North American Practitioner* 4:179–180.

———. 1906. Some forms of hereditary cataract. *Ophthalmic Record* 15:142–151.

Woodruff, C. E. 1900. The alleged breeding of tail-less mice through the inheritance of mutilations. *Medical Record* 58:66–67.

———. 1907. Prevention of degeneration: The only practical eugenics. *Proceedings of the American Breeders Association* 3:247–252.

Woods, F. A. 1908. Recent studies in human heredity. *American Naturalist* 42:685.

———. 1910. Discussion. *Bulletin of the American Academy of Medicine* 11:684–685.

Woods, F. A., Meyer, A., and Davenport, C. B. 1914. Studies in human heredity. *Journal of Heredity* 5:547–555.

Wright, A. R. 1880–81. Hereditary evils. *Transactions of the Homeopathic Medical Society of New York* 16:135–143.

Wright, H. A. 1902. A study of heredity. *Philadelphia Medical Journal* 9:1071–1075.

Wright, H. W. 1914. The function of the general practitioner in relation to the study and prevention of nervous and mental disease. *California State Journal of Medicine* 12:417–421.

Wright, J. 1909. Theories and problems of heredity. *New York Medical Journal* 89:49–53, 309–314, 673–678.

Wyman, M. 1863. The reality and certainty of medicine. *Medical Communications of the Massachusetts Medical Society* 10:234, 237, 239, 250.

Wymer, J. J. 1908–9. Congenital ichthyosis. *New Orleans Medical and Surgical Journal* 61:348–353.

Yoshikawa, T., Rae, V., and Bruins-Slot, W. 1990. Susceptibility to effects of UVB radiation on induction of contact hypersensitivity as a risk factor for skin cancer in humans. *Journal of Investigative Dermatology* 95:530–536.

Young, A. A. 1910–11. Ichthyosis. *Buffalo Medical Journal* 66:129–135.

Young, J. K. 1916. The etiology of congenital absence of parts. *Cincinnati Lancet Clinic* 115:248–250.

Zenner, P. 1894. Friedreich's disease: Hereditary ataxia. *Ohio Medical Journal* 5:177–178.

Ziegler, S. L. 1915. Hereditary posterior polar cataract with report of a pedigree. *Transactions of the American Ophthalmological Society* 14:356–363.

Zirkle, C. 1946. The early history of the idea of the inheritance of acquired characters and of pangenesis. *Transactions of the American Philosophical Society* 35:91–151.

INDEX

A page number followed by *t* designates a table citation.

Library of Congress Cataloging-in-Publication Data

Rushton, Alan R.
 Genetics and medicine in the United States, 1800 to 1922 / Alan R. Rushton.
  p. cm.
 Includes bibliographical references and index.
 ISBN 0-8018-4781-8 (hc : alk. paper)
 1. Medical genetics—United States—History. 2. Human genetics—United
States—History. I. Title.
 [DNLM: 1. Genetics, Medical—history—United States. QZ 11 AA1 R9g
1994]
 RB155.R87 1994
 616′.042′0973—dc20
 DNLM/DLC
 for Library of Congress   93-35943